Apple Pro Training Series

Getting Started with Final Cut Server

Matthew Geller

Apple Certified

Apple Pro Training Series: Getting Started with Final Cut Server
Matthew Geller

Published by Peachpit Press. For information on Peachpit Press books, contact:

Peachpit Press
1249 Eighth Street
Berkeley, CA 94710
(510) 524-2178
Fax: (510) 524-2221
www.peachpit.com

To report errors, please send a note to errata@peachpit.com
Peachpit Press is a division of Pearson Education

Editor: Justine Withers
Series Editor: Nancy Peterson
Production Coordinator: Danielle Foster
Contributing Writer: Jon Rutherford
Technical Editors: Aidria Astravas, Matt McManus, Jon Rutherford, Drew Tucker
Copy Editor: Darren Meiss, Elissa Rabellino
Indexer: Valerie Perry
Cover Illustration: Kent Oberheu
Cover Production: Happenstance Type-O-Rama

ISBN 13: 978-0-321-51024-2
ISBN 10: 0-321-51024-0

9 8 7 6 5 4 3 2 1

Printed and bound in the United States of America

Table of Contents

Getting Started

Welcome to the official training course for Final Cut Server. This book is intended to get both end users and administrators up to speed on the functionality and potential of Final Cut Server.

The Methodology

This book is divided into four sections. The section that will be most useful to you depends on your role within your organization.

- Using Final Cut Server—Lessons 1–6

  This section is intended for end users of Final Cut Server: those who will be working with the software, day to day, within the organization.

- Basic Final Cut Server Administration—Lesson 7

  If you're administering Final Cut Server for the first time, this section will show you how to get started.

- Advanced Final Cut Server Administration—Lessons 8–9

  For administrators looking to extend the capabilities of their Final Cut Server, this section covers advanced administration and workflow techniques.

- Installing and Best Practices—Appendix A

  If you've picked up this book before installing Final Cut Server, this appendix will recommend best practices for installation, as well as guide you through installing the software on the Mac that will be the Final Cut Server.

System Requirements

Before using *Apple Pro Training Series: Getting Started with Final Cut Server,* you should have a working knowledge of your Macintosh and the Mac OS X operating system. Make sure that you know how to use the mouse and standard menus and commands, and also how to open, save, and close files. If you need to review these techniques, see the printed or online documentation included with your system.

Minimum and recommended requirements for the Mac that runs Final Cut Server are outlined in the "Before You Install Final Cut Server" document found on the Final Cut Server installation CD.

About the Apple Pro Training Series

Apple Pro Training Series: Getting Started with Final Cut Server is part of the official training series for Apple Pro applications developed by experts in the field. The lessons are designed to let you learn at your own pace. If you're new to Final Cut Server, you'll learn the fundamental concepts and features you'll need to master the program. If you've recently installed Final Cut Server and want to expand its capabilities, you'll find lessons specifically suited to helping you do so.

Apple Pro Certification Program

The Apple Pro Training and Certification Program is designed to keep you at the forefront of Apple's digital media technology while giving you a competitive edge in today's ever-changing job market. Whether you're an editor, graphic designer, sound designer, special effects artist, or teacher, these training tools are meant to help you expand your skills.

For Final Cut Server administrators completing the course material in this book, you can become an Apple Pro by taking the certification exam at an Apple Authorized Training Center. Certification is offered in Aperture, Color, Final Cut Express, Final Cut Pro, DVD Studio Pro, Logic Pro, Motion, Soundtrack Pro, and Final Cut Server Administration. Certification as an Apple Pro gives you official recognition of your knowledge of Apple's professional applications while allowing you to market yourself to employers and clients as a skilled, pro-level user of Apple products.

To find an Authorized Training Center near you, go to www.apple.com/software/pro/training.

For those who prefer to learn in an instructor-led setting, Apple also offers training courses at Apple Authorized Training Centers worldwide. These courses, which use the Apple Pro Training Series books as their curriculum, are taught by Apple Certified Trainers who balance concepts and lectures with hands-on labs and exercises. Apple Authorized Training Centers have been carefully selected and have met Apple's highest standards in all areas, including facilities, instructors, course delivery, and infrastructure. The goal of the program is to offer Apple customers, from beginners to the most seasoned professionals, the highest-quality training experience.

Resources

Apple Pro Training Series: Getting Started with Final Cut Server is not intended as a comprehensive reference manual, nor does it replace the documentation that comes with the application. For comprehensive information about program features, refer to these resources:

- *Final Cut Server User Manual,* included with your Final Cut Server software, is a comprehensive guide to using Final Cut Server day to day.
- *Final Cut Server Setup and Administration Guide,* included with your Final Cut Server software, explains how to install, configure, and administer Final Cut Server.
- Final Cut Server (www.apple.com/finalcutserver)—Overview of Final Cut Server features and capabilities.
- Final Cut Server Support (www.apple.com/support/finalcutserver)—Support for Final Cut Server.
- Final Cut Server Discussions (http://discussions.apple.com/finalcutserver)—An online forum sponsored by Apple. Users discuss issues and solutions related to Final Cut Server.

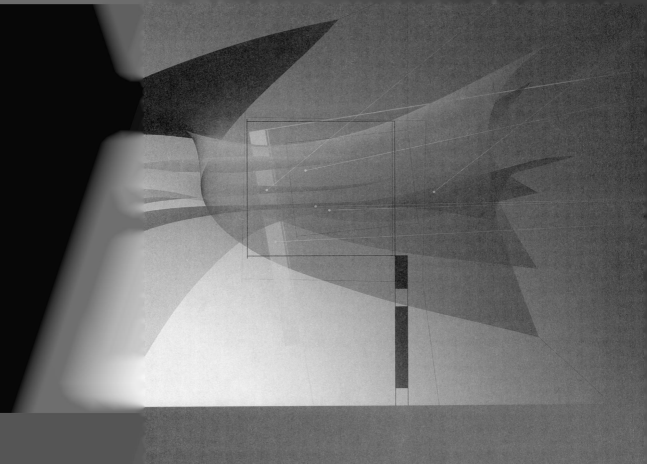

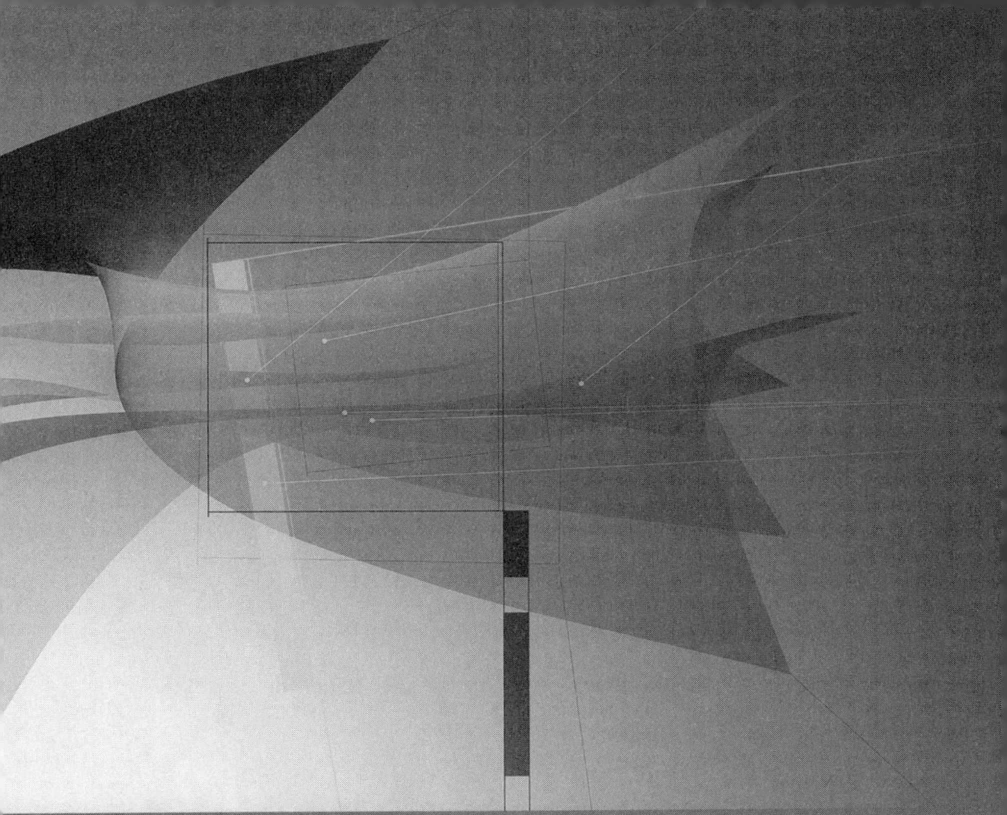

Using Final Cut Server

1

Goals

Understand Final Cut Server terms

Install the client application on Macs or PCs

Tour the client application interface

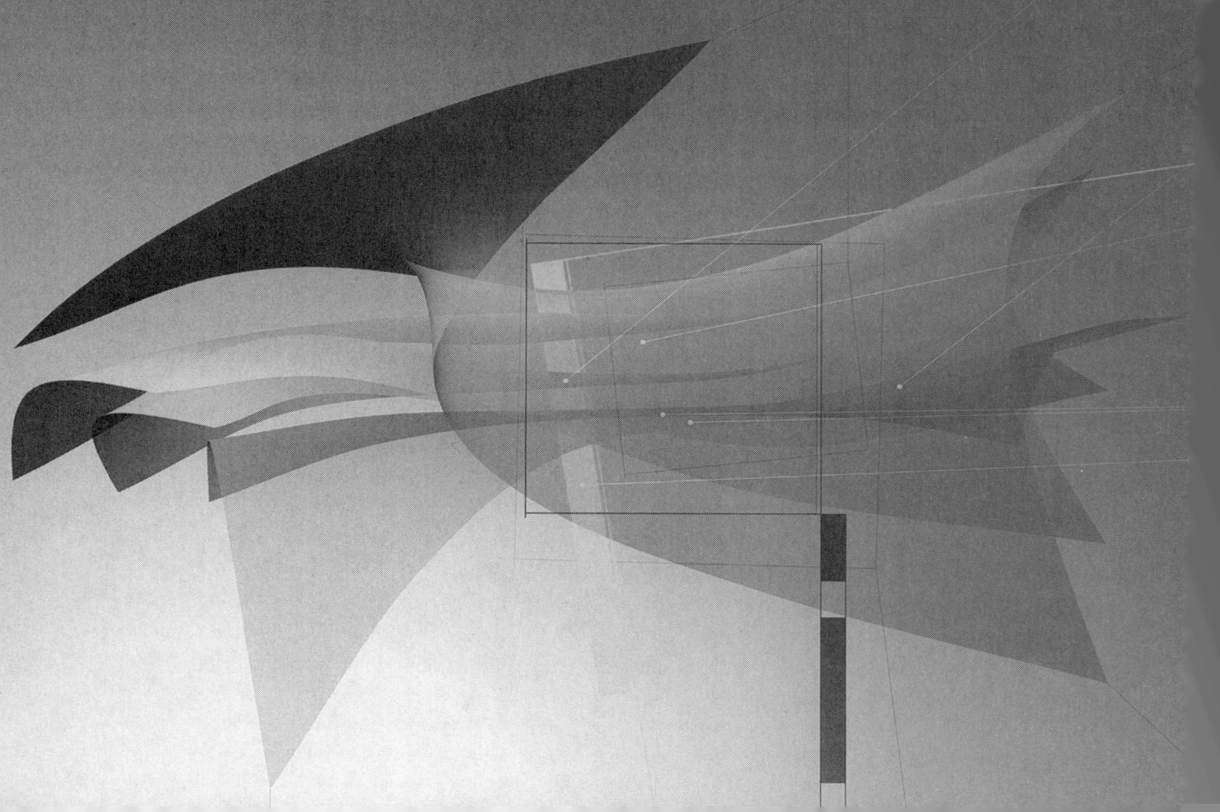

Lesson **1**

Overview and Interface Basics

With Final Cut Server, Apple provides an essential and long-sought-after solution for the collaborative requirements of digital creative work. Final Cut Server has a powerful database and easy-to-use interface that allow it to meet three important needs:

- It manages assets by interacting with both creative application software, most notably Final Cut Pro, and the storage systems of an organization.
- For users inside or outside an organization, it acts as a conduit to complete collaborative steps such as reviewing shots, annotating sequences, and checking files in and out.
- It automates workflow as it generates proxy (low-resolution) versions of clips for quick access; copies and transcodes assets from one storage system to another; scans for new and changed media; and looks for changes in asset metadata, triggering email notifications and a host of other automated actions.

The Nature of the Software: Server, Client, and Device

Final Cut Server gets its name from its nature: It is *server* software that runs on a Mac somewhere in your organization. However, the main way to interact with the software is through a *client* application on any other computer, inside or outside the organization. This client application was developed in the Java programming language and can run on a Mac or PC that is connected to your network.

You'll use the client application to work with all of the files within Final Cut Server, including placing new files inside it and transferring files from it into the applications you use often, like Final Cut Pro.

Your administrator has connected Final Cut Server to any number of *devices* containing media and other files that are important to your staff and your workflow. For Final Cut Server to work properly, any place where your organization stores files should be defined as a device. Here are some device examples:

- An Xsan volume
- A network-accessible file server
- Disk drives that are directly connected to Final Cut Server

The files on these devices are accessible through the network that connects your computer to Final Cut Server. The speed of that connection, whether it is contained within your organization or handled remotely by a virtual private network (VPN), will determine how fast you can access these files.

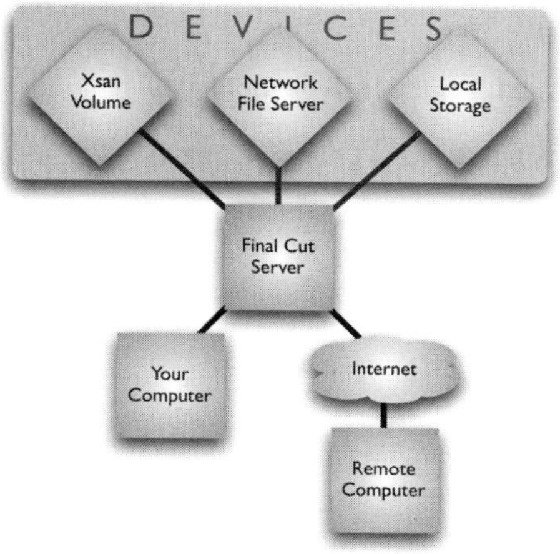

Final Cut Server is the gateway between you and all the places where files reside in your organization. These are called devices.

If your computer and Final Cut Server are Xsan clients accessing the same Xsan volume, Final Cut Server will simply provide pointers to the files on that volume, enabling you to access the files instantaneously. Even in this case, Final Cut Server can also act as a gateway to other devices in your organization.

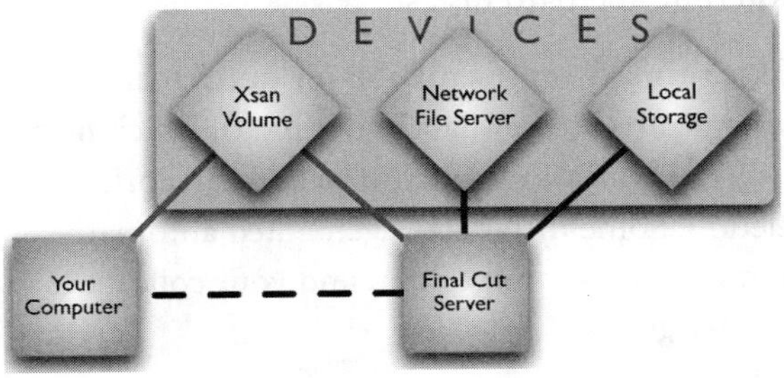

Final Cut Server is used to search for assets on the Xsan volume while it remains the gateway for other devices.

A Few Words on Nomenclature

To avoid confusion when referring to the various manifestations of Final Cut Server terminology throughout:

- *Final Cut Server,* used alone, refers to the Mac that is running installed and configured Final Cut Server software.

  It may also refer to the interaction between the Final Cut Server and the client application in general.
- *Final Cut Server software* refers to the software that gets installed and configured on the Mac that will become the Final Cut Server.
- *Final Cut Server client application,* or just *client application,* refers to the Java client software that users download from the Final Cut Server to install and run on any computer that wishes to communicate with the Final Cut Server.

The Building Blocks: Assets, Proxies, Metadata, Productions, and Jobs

Final Cut Server contains a central catalog of *assets,* which are references to actual media files located on a device. Each asset contains pointers to the original high-resolution version of the file, which we'll call its *primary representation.*

When an asset is added to Final Cut Server's catalog, it makes a *proxy* of the primary representation—a much smaller, lower-quality (and therefore easier-to-transmit) version of the primary representation that is usually stored on the hard disk(s) of Final Cut Server. Users that connect to Final Cut Server using low-bandwidth connections will view and use proxy files in order to save transmission time and hard disk space.

Each asset contains a *thumbnail* image of the primary representation for quick identification. The asset also contains a rich assortment of *metadata* (data about data), which helps you in searching for the asset, using it in your organization's workflow, and categorizing it for special tasks such as archiving or deletion. Some metadata is generated automatically when the asset is created, but a lot of metadata is input by you and your colleagues in order to document the asset's attributes and usage.

Assets can be pooled together in a group called a *production*. A production is like a virtual folder inside of Final Cut Server: a group of assets being used for the same purpose. Final Cut Server can assign metadata to productions automatically, and users can add metadata to them. Productions have their own catalogs within Final Cut Server that can be searched and categorized.

Final Cut Server also tracks *jobs*—the actions you, your colleagues, and Final Cut Server itself perform, such as adding assets to a catalog or copying original representations from one device to another. All jobs are logged and can be searched by users. Jobs have certain metadata attached to them, including error messages that can be looked up in case a job fails.

Collaboration: Users and Groups

Obviously, Final Cut Server will be accessed by *users*, people inside and perhaps outside the organization. In order to gain access to the catalogs within Final Cut Server, you will log in to it with a username and a password. This provides security for assets and productions that have sensitive materials. Your administrator gathers certain users together into *groups*. These groups will have certain access and functionality privileges within Final Cut Server, so that everyone has access to the appropriate assets and functions.

Downloading the Client Application

The requirements for the Final Cut Server client application are simple: It must run on a Windows XP, Windows Vista, or Mac OS X system on which a runtime version of Java and Apple's QuickTime are installed. Java and QuickTime come preinstalled on Mac OS X. Some PCs may require the installation of Java and QuickTime before you can install the Final Cut Server client application. Contact your administrator to install Java or QuickTime.

> **MORE INFO ►** These instructions are for end users wanting to download and install the client application on their computers. For system requirements and instructions on how to install and configure Final Cut Server, please refer to the Appendix and Setup and Administration Guide that came with your software.

Your administrator may have already installed the Final Cut Server client application on your Mac or PC. If not, it is always available for download via a web browser at the network address, or URL, of the Final Cut Server.

The download URL can be either a specific domain name or an IP address, followed by */finalcutserver*.

Some examples:

Your administrator can furnish this URL.

> **NOTE ►** This web page is only for downloading the Final Cut Server client application. Once you have downloaded and installed the client application, you will start it on your own machine.

When you go to this URL, you will first see a page that prompts you to download the Final Cut Server client application.

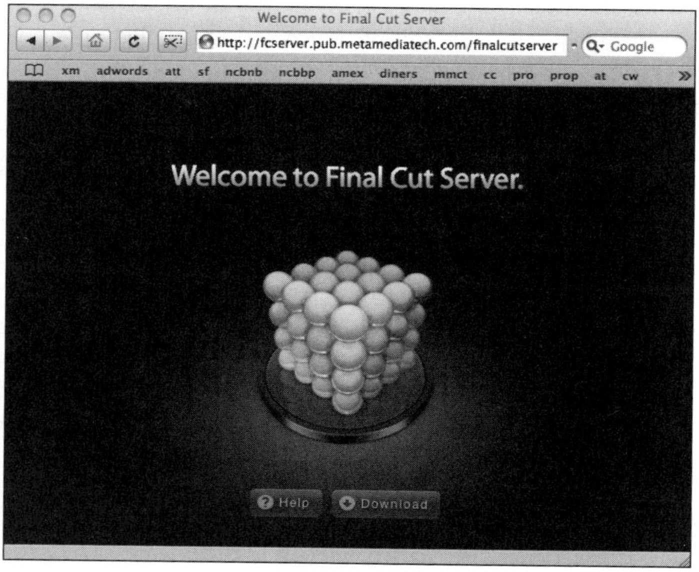

Click Download. If you are accessing the Final Cut Server software through a VPN outside of the organization, this download may take some time.

The next few steps differ slightly in how the application downloads on a Mac or a PC, so we've included instructions for both.

Installation Instructions for Mac

Once you click Download, you will see a progress indicator window showing that the software is downloading.

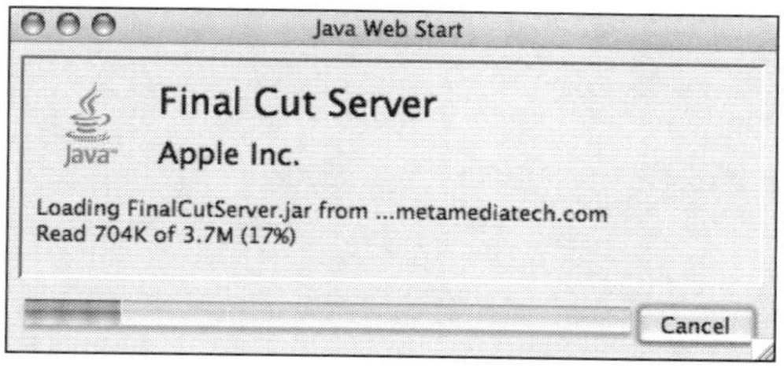

1 When the download is complete, you will tell your computer that the application it just downloaded should be trusted. In the next dialog, click Show Certificate.

http://localhost
This application was signed by "Apple Inc.," but Java cannot verify the authenticity of the signature's certificate. Do you trust this certificate?
Click Trust to use this application and allow it unrestricted access to your computer. Click Don't Trust to quit this application.
Show Certificate
Don't Trust
Trust

2 The dialog will expand. Select the checkbox next to the statement "Always trust 'Apple Inc.'"

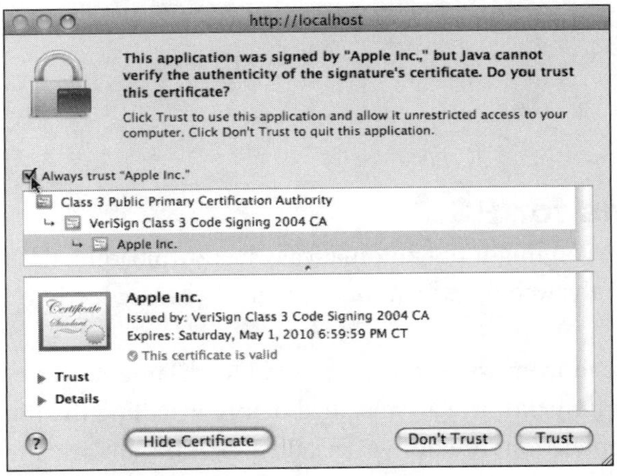
http://localhost
This application was signed by "Apple Inc.," but Java cannot verify the authenticity of the signature's certificate. Do you trust this certificate?
Click Trust to use this application and allow it unrestricted access to your computer. Click Don't Trust to quit this application.
Always trust "Apple Inc."
Class 3 Public Primary Certification Authority
VeriSign Class 3 Code Signing 2004 CA
Apple Inc.
Apple Inc.
Issued by: VeriSign Class 3 Code Signing 2004 CA
Expires: Saturday, May 1, 2010 6:59:59 PM CT
This certificate is valid
Trust
Details
Hide Certificate
Don't Trust
Trust

3 Click the Trust button at the bottom of the dialog.

4 Once the client application is accepted, you will be asked if you'd like to make a desktop shortcut. Click Yes to do so.

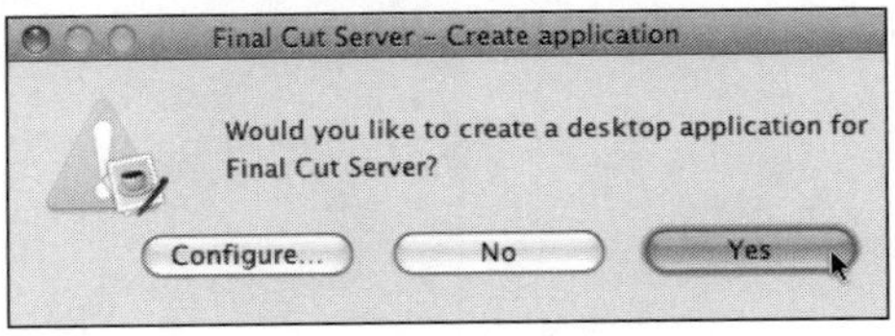
Final Cut Server – Create application
Would you like to create a desktop application for Final Cut Server?
Configure...
No
Yes

5 The final dialog allows you to choose where you'd like to save the application.

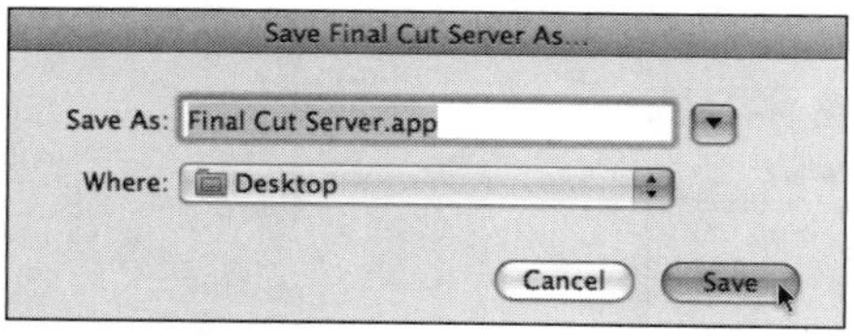

From now on, you can simply double-click the Final Cut Server application shortcut on your desktop. You can also drag this icon to the Dock.

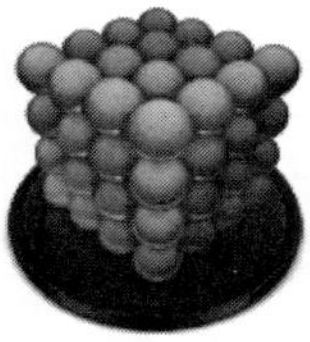

Installation Instructions for PCs

Depending on your security settings, you might not automatically be downloading the software once you click Download from the web page mentioned in the preceding section.

1 Be sure that Java (a free download from www.java.com) and QuickTime (a free download from www.apple.com/quicktime/) are installed before installing the Final Cut Server software. If these applications are not yet installed, first install Java and then Quicktime, in that order.

2 If you've done that and nothing seems to be happening, locate the Final Cut Server file on your desktop and double-click it. However, if you do see a Java window indicating that the application is downloading, skip on to Step 5.

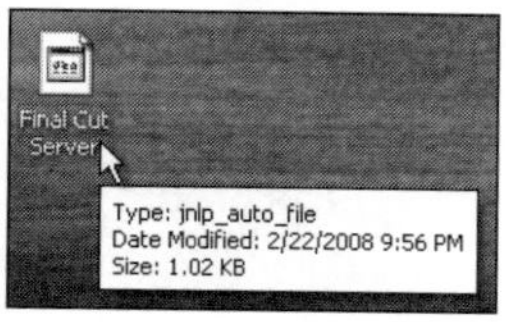

3 The next dialog will ask how to open the file. Click the "Select the program from a list" radio button and click OK.

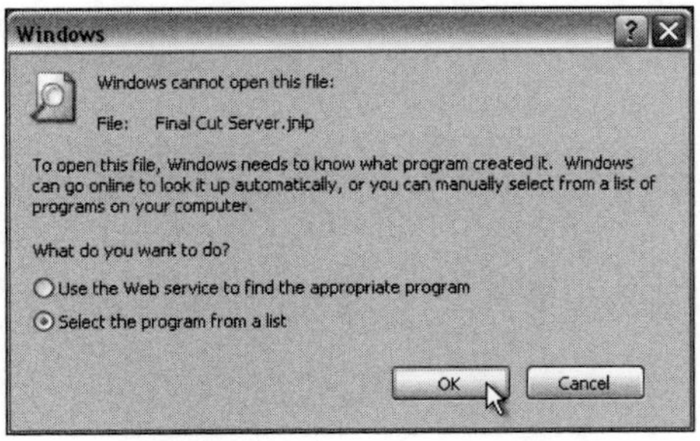

4 In the next dialog, select the Java™ Web Start Launcher application. Make sure to select the checkbox next to "Always use the selected program to open this kind of file" and then click OK.

The program should start downloading.

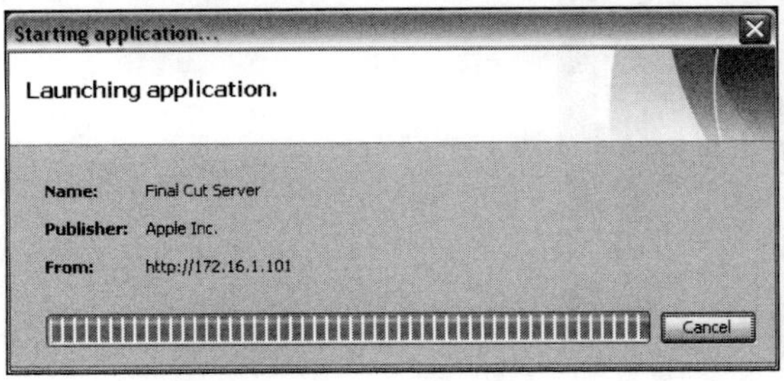

5 In the dialog that opens, select the checkbox next to "Always trust content from this publisher" and click Run.

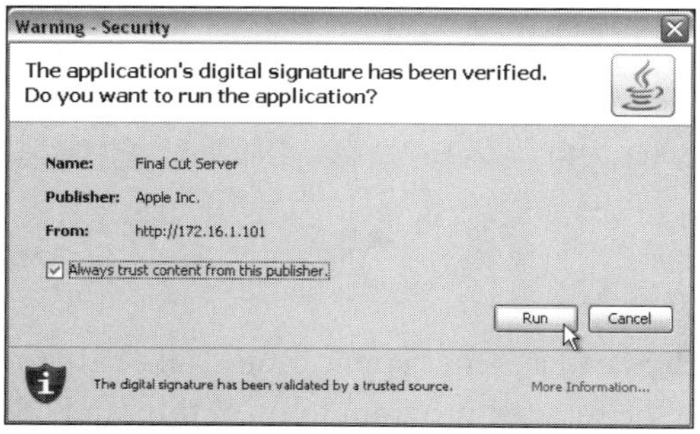

On PCs, the client application will now appear as an icon on your desktop.

It will also be located in the Start Menu under Apple > Final Cut Server.

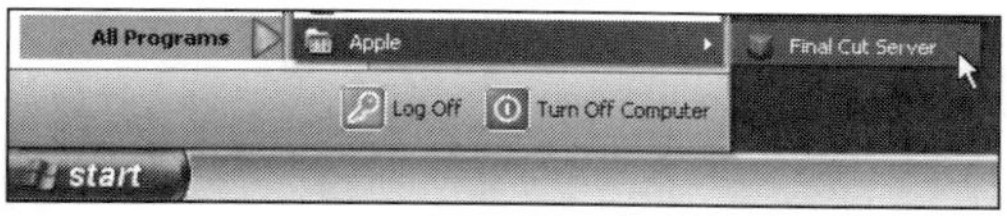

Logging In to Final Cut Server

After you start the client application, the login process is the same for both Mac and PC.

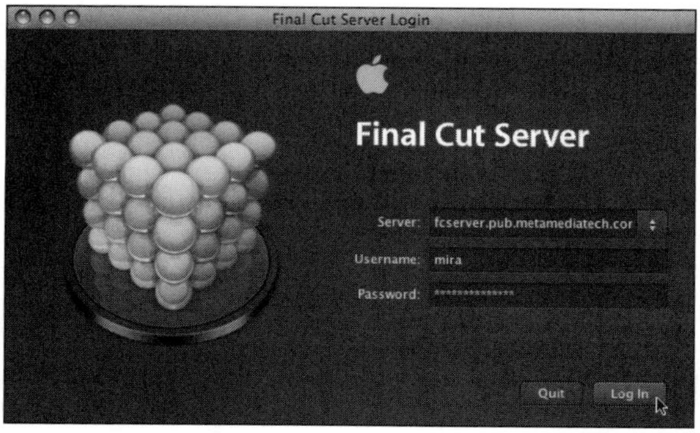

The Final Cut Server administrator will provide you with a username and a password. These might be unique, or they may be the same ones that you use for other services at the organization.

Logging in brings you to the main interface.

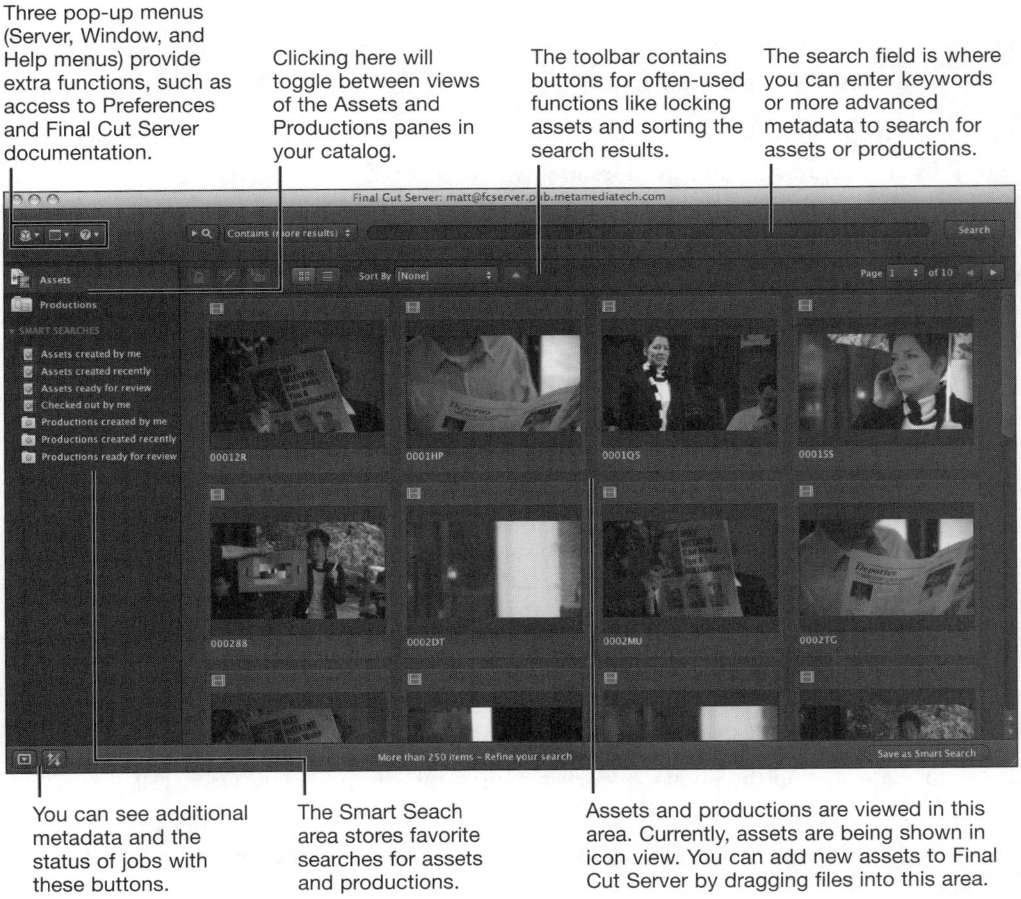

Three pop-up menus (Server, Window, and Help menus) provide extra functions, such as access to Preferences and Final Cut Server documentation.
Clicking here will toggle between views of the Assets and Productions panes in your catalog.
The toolbar contains buttons for often-used functions like locking assets and sorting the search results.
The search field is where you can enter keywords or more advanced metadata to search for assets or productions.
You can see additional metadata and the status of jobs with these buttons.
The Smart Seach area stores favorite searches for assets and productions.
Assets and productions are viewed in this area. Currently, assets are being shown in icon view. You can add new assets to Final Cut Server by dragging files into this area.

In the next five lessons, you will learn how to work with all the features of Final Cut Server's client application.

Lesson 2, "Working with Assets," shows you how to use Final Cut Server to add, search, annotate, and retrieve assets.

Lesson 3, "Working with Final Cut Studio Projects," shows you the additional workflow-related functions specific to working with FCP and other Final Cut Studio project files in Final Cut Server.

Lesson 4, "Working with Productions," shows you how to group assets together into productions to aid your workflow.

Lesson 5, "Working with Devices," shows you how to search devices whose media may not be added as assets to Final Cut Server.

Lesson 6, "Working with Jobs," shows you how to use the Search All Jobs window to see how jobs are progressing and troubleshoot why they might have failed.

Lesson Review

1. What is the difference between a primary representation and a clip proxy?
2. True or false: The Final Cut Server client application can run on Windows Vista.
3. What are the two main panes within Final Cut Server?
4. What is metadata?
5. Where can you access Final Cut Server's Preferences menu within the client?

Answers

1. The primary representation is the original file from which the asset was made. Clip proxies are smaller, lower-resolution files derived from the primary representation of a video asset. They're more easily transmitted, especially over a low-bandwidth connection.
2. True. You can run the Final Cut Server client application on Mac OS X v10.4 or later and Windows XP or Vista operating systems. The computer used as the Final Cut Server needs to have Mac OS X v10.5 Leopard or Mac OS X Server v10.5 Leopard or later installed.

3. The Assets and Productions panes, which can be accessed from the main client application window in the upper-left corner.
4. *Metadata* means data about data. It is used to describe, find, and repurpose content. Final Cut Server has powerful search fields where you can search for assets and productions using metadata information.
5. Preferences can be found in the Server pop-up menu in the upper-left corner of the main client window.

2

Goals

- Perform basic and advanced searches for assets
- View asset details, add metadata, and annotate assets
- Upload assets into Final Cut Server
- Preview and add assets to your cache
- Prepare assets for disconnected use
- Export assets out of Final Cut Server
- Archive, restore, and delete assets
- Duplicate assets to different devices
- Check out, check in, lock, and unlock assets

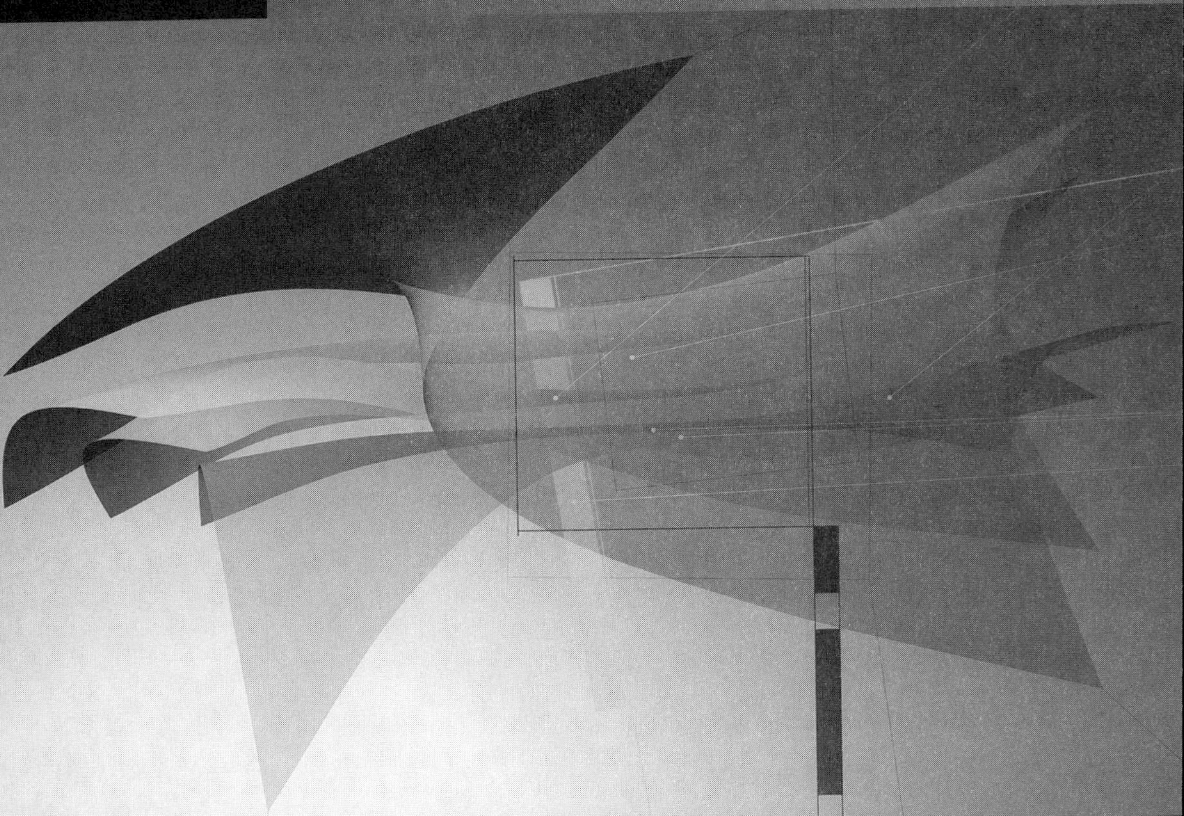

Lesson **2**

Working with Assets

Assets are the building blocks of your organization's Final Cut Server catalog. This lesson shows you how to search, enrich, create, and use the assets contained within Final Cut Server.

Assets in Final Cut Server are symbolic. They refer to actual files called *representations*, which are located on the various devices to which Final Server is connected.

- The *primary representation* is the original file that was used to make the asset. This could also be referred to as the high resolution, or hi-res, media.
- For a clip asset, the *clip proxy* is a video clip that Final Cut Server makes from the primary representation of a video asset. It is much smaller and of lower quality than the original representation, and therefore can be downloaded and viewed quickly, even over an Ethernet network.
- Some organizations will also have *edit proxies*, which use Apple's ProRes 422 codec. Edit proxies are used for editing clips that have original representations that are uncompressed or in different formats. The ProRes 422 codec therefore becomes an equalizer for all these assets, making it perfect for an editing base.

- The *poster frame* is a single image made from any kind of visual asset, and it represents the essence of the content. It's usually in JPEG format.
- The *thumbnail* is a tiny image of any visual asset, and it represents the asset in a search result.

We'll use these terms throughout this lesson to show you how to access and use these asset representations.

Searching for Assets

Searching for assets is one of Final Cut Server's most powerful features. The more specific your search criteria, the more meaningful the search result will be.

Searching is done in the top area of the client application window.

Choose a search method from the pop-up menu.
Type a keyword here.
Click here to perform the search.
Contains (more results)
journey
Search
Click here to clear the search field.

Structuring Search Queries

There are two selections in the Search Type pop-up menu: Contains and Matches Word. In either case, Final Cut Server searches through all metadata for your keywords. Don't worry about case sensitivity, because it is ignored when searching.

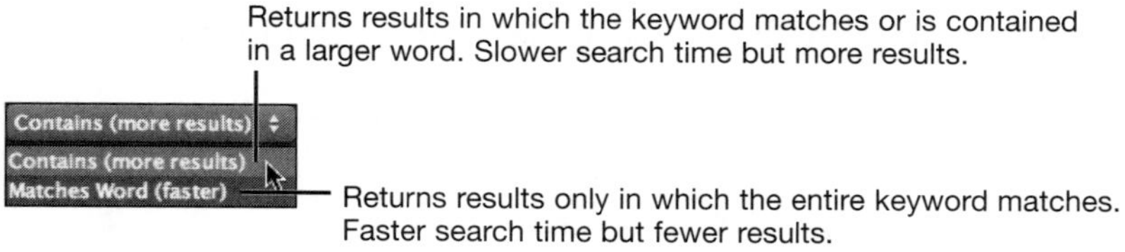
Returns results in which the keyword matches or is contained in a larger word. Slower search time but more results.
Contains (more results)
Contains (more results)
Matches Word (faster)
Returns results only in which the entire keyword matches. Faster search time but fewer results.

You can search for multiple keywords by typing them together, separated by spaces. Assets whose metadata contains all typed keywords will be returned.

journey bridge

You can exclude a keyword with the – sign (hyphen). In the example below, assets whose metadata contains the word *journey* but not the word *bridge* will be returned.

If you need to match an exact phrase, place it in quotes and choose the "Matches Word (faster)" option from the pop-up menu to the left. Here, only assets whose metadata matches the exact phrase will be returned.

Selecting Advanced Search Options

The advanced search options allow you to refine your search even further, by filtering for specific metadata fields.

Click here to reveal and hide the advanced search options.

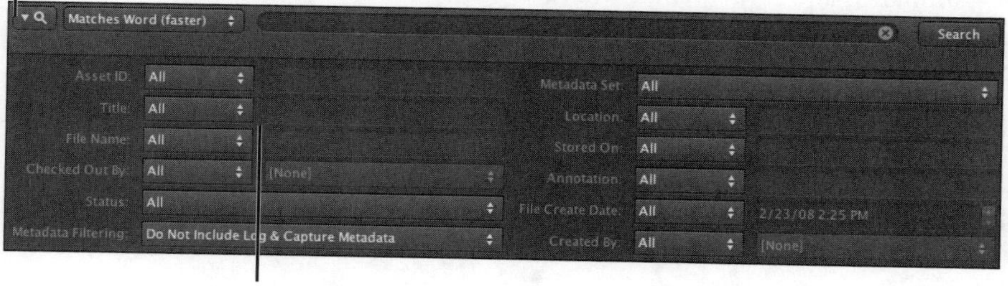

Use pop-up menus to further refine an advanced search.

You can search any or all fields that are shown in the advanced search options, in addition to keyword searches in the main search field.

For example, you can search for assets checked out by a particular user by using the Checked Out By field. Narrowing down search results to just video clips is accomplished using the Metadata Set field.

Additionally, there are pop-up menus to the right of each field name that allow you to refine the search. Pop-up menu options vary by field, but two major ones are for numbers and text.

> **NOTE ▸** If your administrator has customized the layout of the advanced search options, you may see more or fewer fields than are pictured here. If your organization requires additional search fields, talk to your administrator about customizing the advanced search options.

The following modifiers are available for text-based fields, such as Title, Location, Stored On or Annotation:

- All—this field will be ignored
- Equals—exact match to text
- Not Equals—match anything but the text
- Not Equal and Not Blank—match anything but the text only when the field has data in it
- Contains—contains the text
- Begins with—text begins the field
- Ends with—text ends the field
- Matches Word—exact match to text, using words to determine matches rather than other characters
- Any Of—similar to an "or" search, it will return at least one word used in the search

The following modifiers are available for numerical fields, such as Asset ID:

- All—this field will be ignored
- = —equal to
- < —less than
- > —greater than
- <= —less than or equal to
- >= —greater than or equal to
- != —does not equal

> **NOTE ▸** To search Final Cut Pro Log & Capture metadata, such as log notes and source timecode from Final Cut Pro projects, activate the advanced search option and set the Metadata Filtering pop-up menu to Include or Only Include Log & Capture Metadata.

Viewing Search Results

Search results will appear at the bottom of the window. You can change how you view the results with the Toolbar.

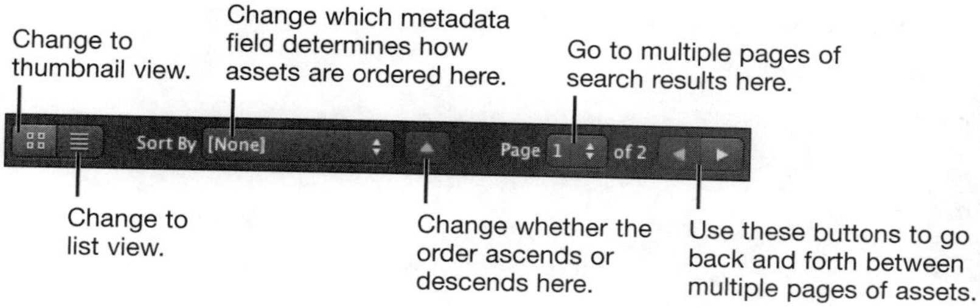
Change to thumbnail view.
Change which metadata field determines how assets are ordered here.
Go to multiple pages of search results here.
Sort By [None]
Page 1 of 2
Change to list view.
Change whether the order ascends or descends here.
Use these buttons to go back and forth between multiple pages of assets.

Your search might return more results than can be seen in one page. You can use the page-navigation controls in the upper-right corner of the window to view additional pages of results.

You can also view additional details about an asset with the Show Item Information button in the lower left of the window.

In all views, you will see an identifying icon that visually denotes the kind of asset you are viewing. Clicking this icon when viewing a search result in thumbnail view allows you to preview visual assets, such as images and video clips.

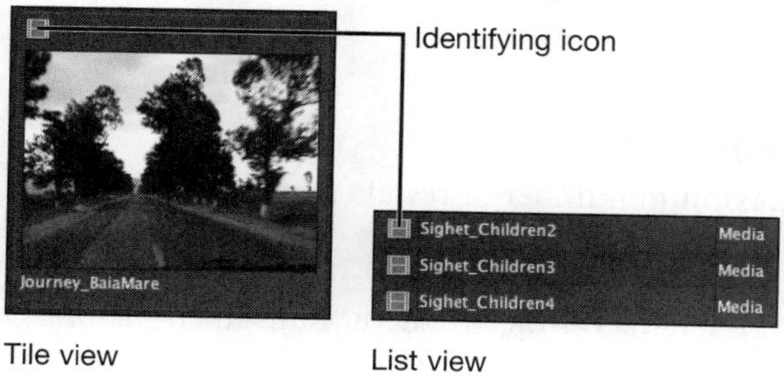
Identifying icon
Journey_BaiaMare
Sighet_Children2
Media
Sighet_Children3
Media
Sighet_Children4
Media
Tile view
List view

NOTE ► You can only preview visual assets, such as images and video clips, while viewing a search result in thumbnail view, not list view. Any other kind of asset, such as audio clips, cannot be previewed. You must add them to your cache to use them, as explained later in this lesson.

Final Cut Server assigns the following icons to assets depending on their type:

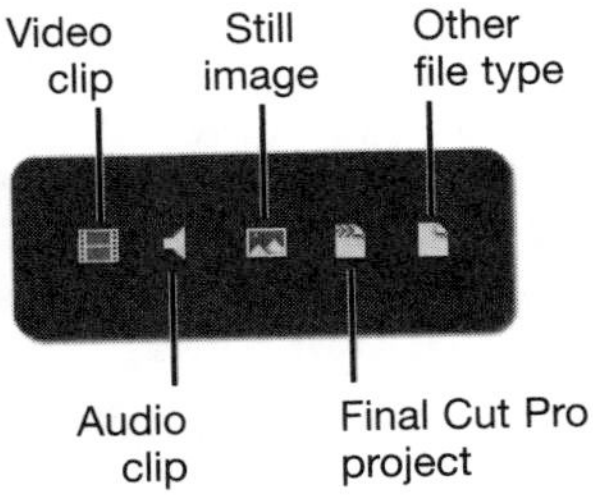

In some cases, you may see one or more additional icons, to the right of the identifying icon, which will visually describe the state of the asset.

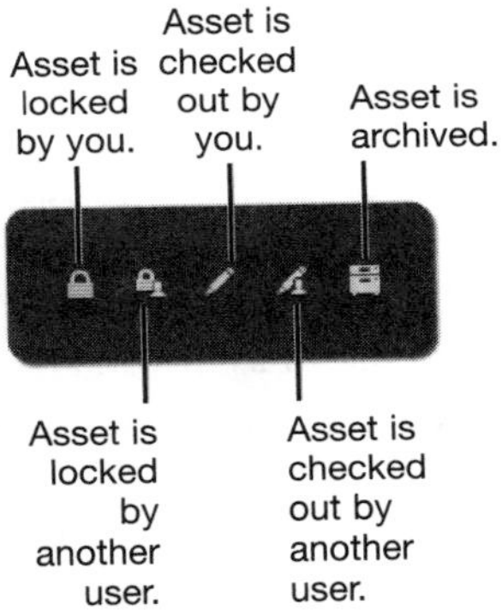

Adjusting Search Results Settings

The client application has limits on the maximum number of results that are returned and the maximum number of results per page.

To change these and other settings, select Preferences from the Server pull-down menu in the upper-left area of the client application window.

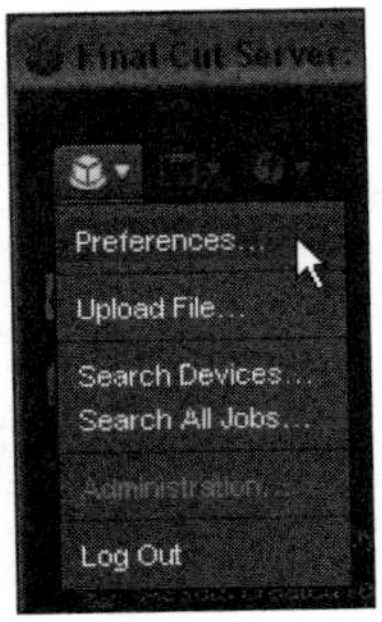

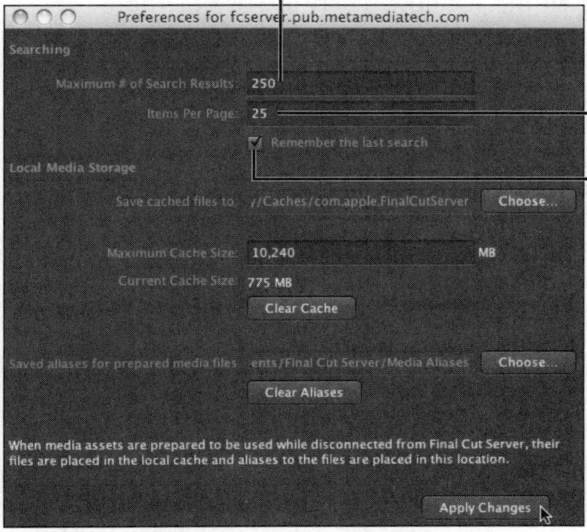
Change the maximum number of returned results.
Change the number of items on a page.
With “Remember the last search” checked, every time you log in, the client application will display the results of the last search you made.
Preferences for fcserver.pub.metamediatech.com
Searching
Maximum # of Search Results: 250
Items Per Page: 25
Remember the last search
Local Media Storage
Save cached files to: //Caches/com.apple.FinalCutServer
Choose...
Maximum Cache Size: 10,240 MB
Current Cache Size: 775 MB
Clear Cache
Saved aliases for prepared media files: ents/Final Cut Server/Media Aliases
Choose...
Clear Aliases
When media assets are prepared to be used while disconnected from Final Cut Server, their files are placed in the local cache and aliases to the files are placed in this location.
Apply Changes

Working with Smart Searches

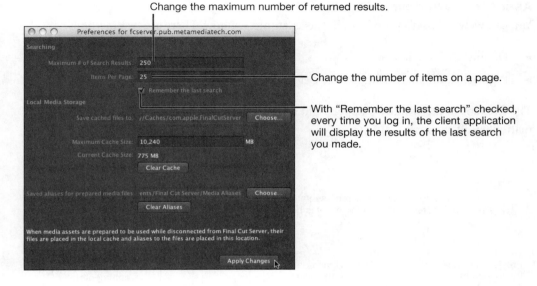

Smart Searches allow you to save simple or advanced search criteria in a single click. When you have fine-tuned a search that you'll often come back to, click the Save as Smart Search button in the lower right of the main client application window.

Save as Smart Search

You'll then be prompted to name the Smart Search on the left side of the main client application window.

New Amy Winehouse clips

You can now easily access the search criteria for that search by clicking its icon.

To delete a Smart Search, Control-click (right-click) the entry and select Delete.

Accessing Asset Options and Properties

When you've found the assets you need, Final Cut Server provides a number of options for working with them.

Asset Info Window

You can view an asset's properties in the asset info window. There are two methods:

- Double-click on an asset.
- Right-click on an asset and select Get Info.

The five tabs in the window's upper-right area open panes that give access to more information about the asset.

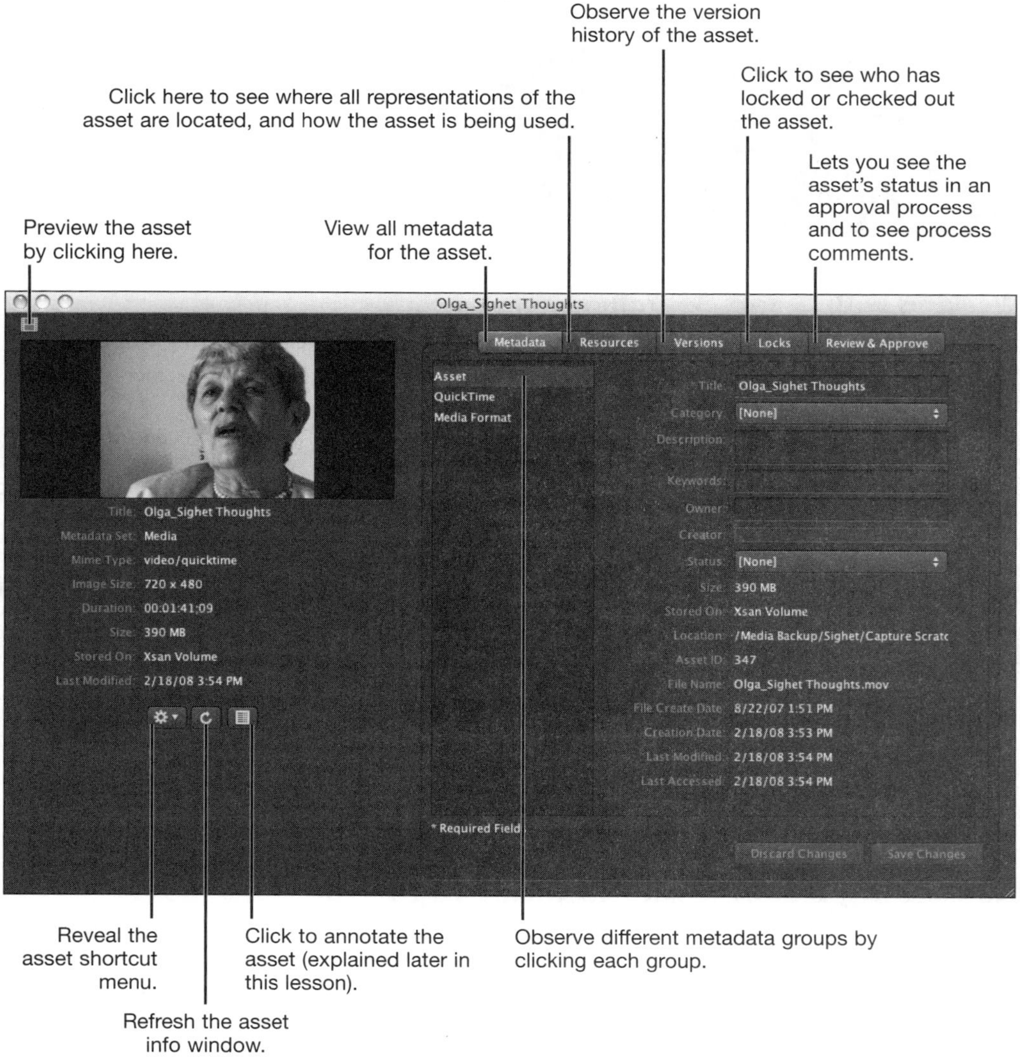

Locking and Unlocking Assets

Assets can be locked so that others cannot delete or overwrite the asset or its primary representation. There are two methods:

- Select Lock from the asset shortcut menu. You can access the asset's shortcut menu by Control-clicking its icon in thumbnail view.
- Select the asset in a search results list, and click the Lock button in the Toolbar in the main client application window.

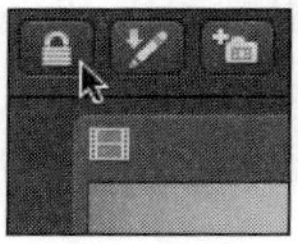

A lock icon will appear in the assets listing (in any view) to indicate that you locked it.

Only you or an administrator can unlock an asset that you originally locked. There are four methods:

- Click the lock icon directly on the asset's icon in a thumbnail view.
- Select Unlock from the asset shortcut menu.
- Select the asset in a search results list, and click the Unlock button in the Toolbar in the main client application window.

- Administrators can clear the lock from the Locks tab in the asset info window.

To see who has locked an asset, click the Locks tab in the asset info window.

> **MORE INFO ►** If you encounter a locked asset that you need to work with, you can still carry out many tasks with it, including exporting and duplicating, explained later in this lesson.

Reanalyzing Assets

Occasionally, assets may not have been analyzed properly when first added to the Final Cut Server catalog. Errors in the creation of a clip proxy, poster frame, or thumbnail, or simply changes to the original representation that were not tracked in Final Cut Server, call for a reanalysis of the primary representation. If this is necessary, you can ask your administrator to re-analyze the asset.

Once you administrator logs in, he or she can choose Analyze from the asset shortcut menu. Analyzing an asset re-creates its thumbnail, poster frame, and clip proxy (in the case of a clip asset).

> **NOTE ►** For administrators: If analysis of an asset repeatedly fails, look at the job window in Final Cut Server to understand why it's not working. You can find out more about the job window in Lesson 6, "Working with Jobs."

Enriching Assets with Metadata and Annotation

Assets become more valuable to your organization as their metadata is customized. Annotation adds more specific information to particular timecode locations in a video clip asset.

Adding and Editing Asset Metadata

Adding and editing metadata is done in the asset info window. The basic rule is: Any metadata field that accepts text from your keyboard can be modified with custom metadata.

Description:
Olga desrcibes her life as a girl in Sighet.

If a field doesn't accept keyboard text, it usually means that the field has been restricted for use by Final Cut Server, your administrator, or users with higher access permissions than yours.

Pop-up menus are also ways to customize the metadata. If you see them instead of text-entry fields, your administrator wants to use only preselected words or phrases to describe a particular metadata attribute.

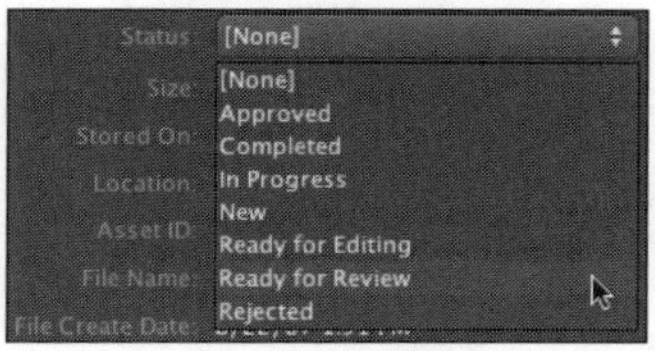

Use of pop-up metadata keywords or phrases ensures consistent search terms for an entire organization.

After adding or editing the metadata, make sure to click the Save Changes button in the lower right.

If you close the window or switch to another tab of the window without saving your changes, Final Cut Server will remind you of your changed metadata and prompt you to save.

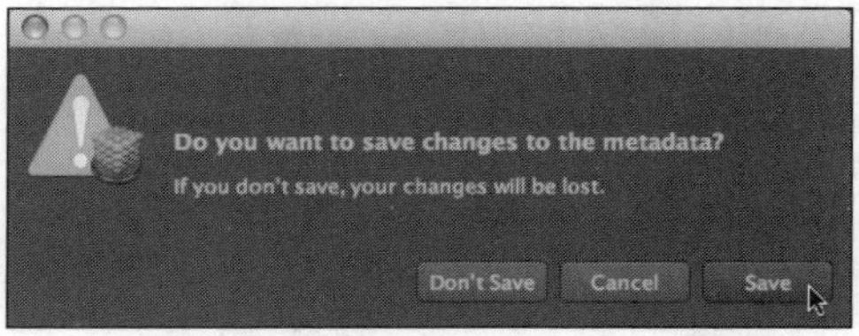

Annotating Clips

Video clip assets can be annotated with comments that can be viewed by others in your organization. Annotations can describe an entire asset or refer to a particular section specified by In and Out points.

For example, annotations can be used to show what parts of a clip should be used within an edit, or to indicate issues with the footage.

Clicking the Annotate button in the asset info window brings up the annotations window.

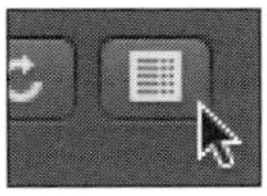

The clip proxy is used to create the annotations. If you are viewing the proxy for the first time, there will be a delay as the clip proxy is added to your computer's cache.

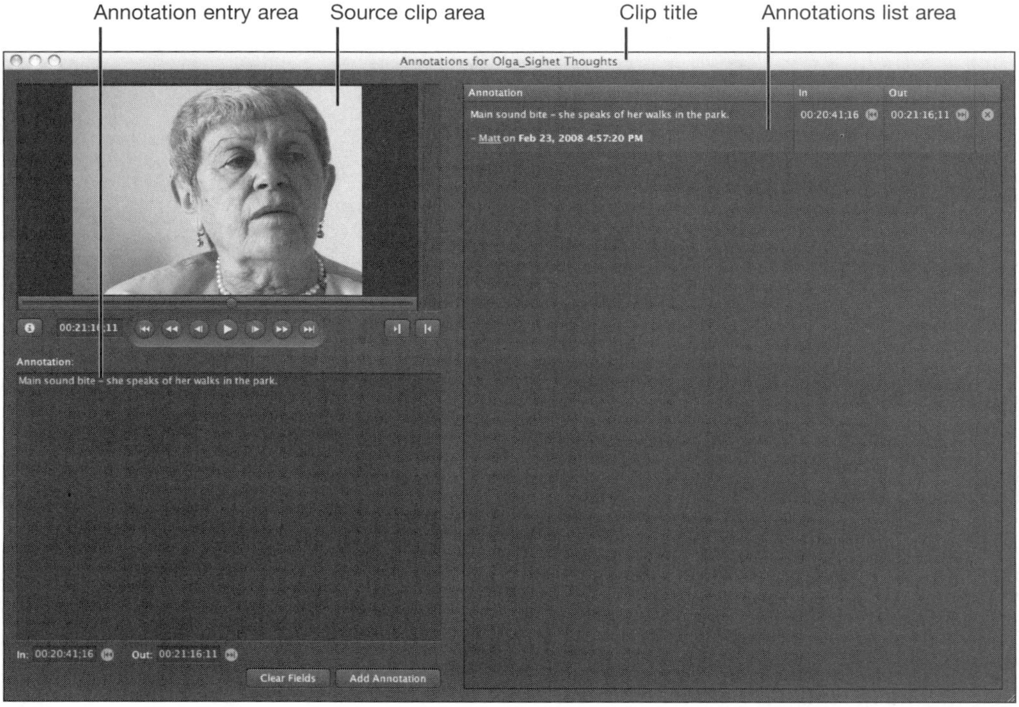

The source clip area contains the source clip and its audio levels, transport controls for playing the media, In and Out marker buttons, a display of the source timecode, and a button to take you back to the asset info window.

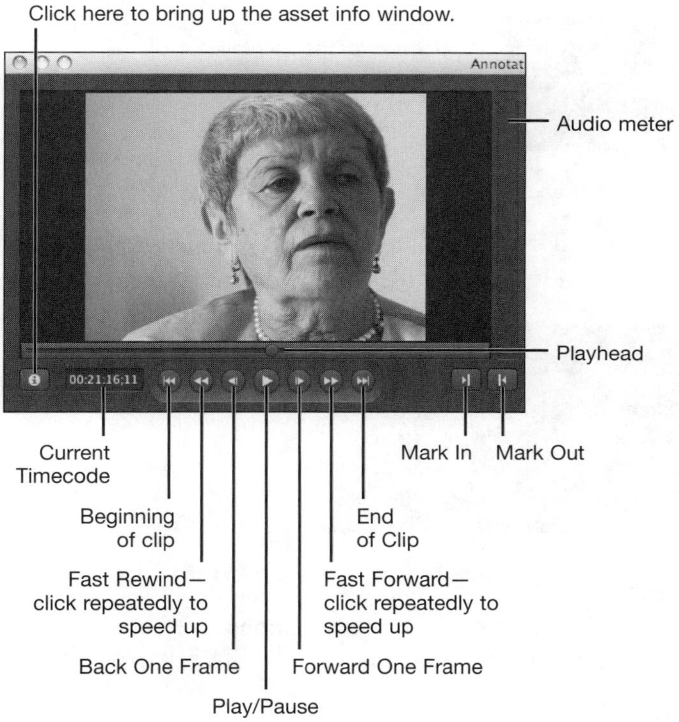

You can also use the Spacebar to start and stop playback, and the Left and Right Arrow keys to advance one frame at a time backward and forward, respectively. Pressing the greater than (<) and less than (>) keys shuttles the clip back and forth, and pressing them multiple times increases the playback speed in that respective direction. Press the I and O keys to set In and Out points, respectively.

The annotation entry area is where you type the annotations for the marks you set in the source clip area.

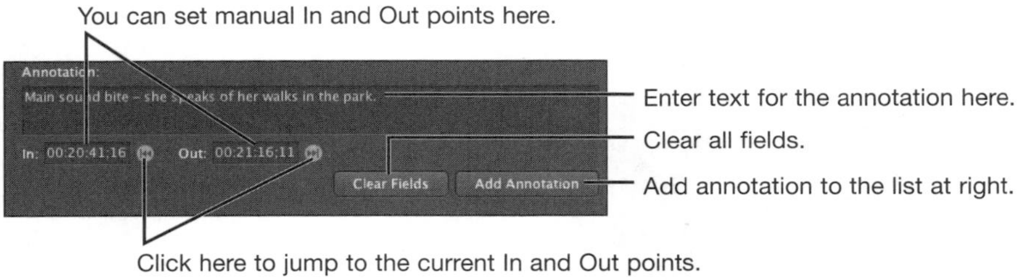
You can set manual In and Out points here.
Annotation:
Main sound bite – she speaks of her walks in the park.
In: 00:20:41:16
Out: 00:21:16:11
Clear Fields
Add Annotation
Enter text for the annotation here.
Clear all fields.
Add annotation to the list at right.
Click here to jump to the current In and Out points.

The annotations list area contains the full list of annotations you have added to the clip. Double-clicking any annotation will load it back into the annotation entry area.

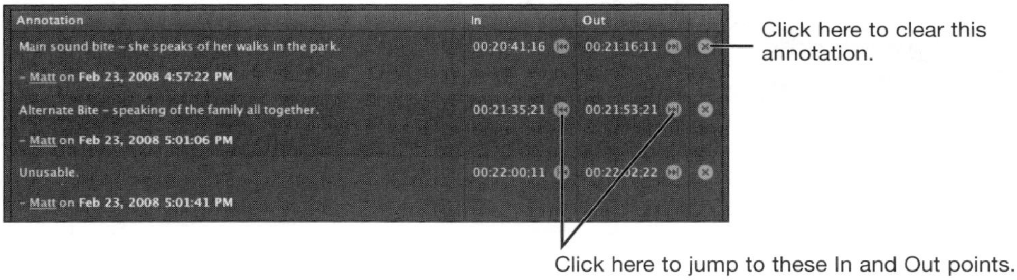
Annotation
In
Out
Main sound bite – she speaks of her walks in the park.
00:20:41:16
00:21:16:11
– Matt on Feb 23, 2008 4:57:22 PM
Alternate Bite – speaking of the family all together.
00:21:35:21
00:21:53:21
– Matt on Feb 23, 2008 5:01:06 PM
Unusable.
00:22:00:11
– Matt on Feb 23, 2008 5:01:41 PM
Click here to clear this annotation.
Click here to jump to these In and Out points.

Close the annotations window when you are finished. Once annotations have been saved for an asset, they can be viewed quickly by selecting Annotations from the asset shortcut menu.

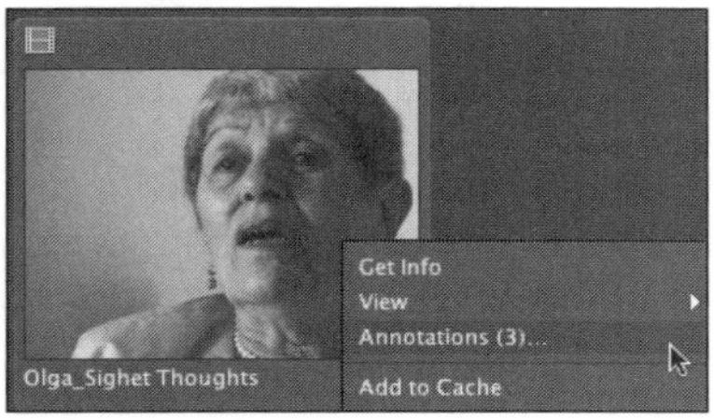
Get Info
View
Annotations (3)...
Add to Cache
Olga_Sighet Thoughts

Getting Assets into Final Cut Server

There are three methods for turning new or modified files into assets in Final Cut Server: Upload, Watchers, and Scanning.

Upload—If the files currently reside only on the hard drive of your computer, use this method. It requires the most interaction from you, but also allows you to customize metadata for the asset that will be created.

Watcher—If you are currently inside your organization's network, either physically or through a VPN, then you may just need to place the files in a watcher, which is explained in more detail later in this section.

Scanning—Files that get created on a shared device such as an Xsan volume may have assets automatically created for them using a method called Scanning, also described later in this section.

Uploading Files

The Upload method is used for three reasons:

- To copy newly created files from your computer to a device that is connected to Final Cut Server.
- To create assets for the files, which will automatically generate the other representations, including clip proxies for video files.
- To have an opportunity to customize the metadata for the assets that will be created for the files.

Upload should therefore be used any time you have created new files locally—on your computer's hard drive—that your organization needs to share. Because the client application communicates with Final Cut Server over a network connection, you can add assets to your organization's devices no matter where you are. Be aware, however, that the speed of your connection to Final Cut Server will determine the speed of the upload of the files.

There are two methods for uploading files:

- Select Upload Files from the Server pull-down menu.
- Drag and drop.

Using the pull-down menu is the best method for selecting a large group of files within a folder. Begin by selecting Upload File from the Server pull-down menu in the upper left of the client application window.

An Upload dialog appears, prompting you to select a file, a series of files, or a folder.

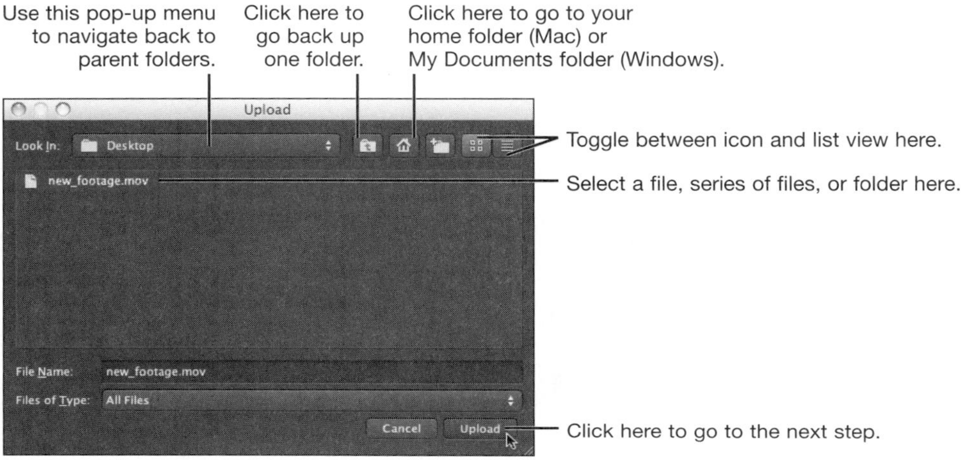
Use this pop-up menu to navigate back to parent folders.
Click here to go back up one folder.
Click here to go to your home folder (Mac) or My Documents folder (Windows).
Upload
Look In:
Desktop
Toggle between icon and list view here.
new_footage.mov
Select a file, series of files, or folder here.
File Name:
new_footage.mov
Files of Type:
All Files
Cancel
Upload
Click here to go to the next step.

However, if you like to use your mouse, using drag and drop couldn't be easier. Simply drag and drop a file, a series of files, or an entire folder into the client application window.

MOVIE
Sighet_Synagogue_Int18
Sighet_Synagogue_
MOVIE
new_footage.mov

If you upload a QuickTime reference movie (which refers to other media content instead of containing all elements within itself), Final Cut Server will advise you to flatten the file before uploading it.

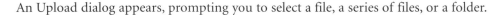

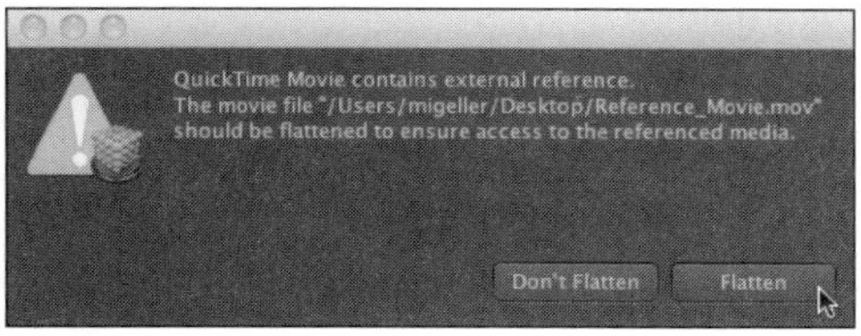
QuickTime Movie contains external reference.
The movie file "/Users/migeller/Desktop/Reference_Movie.mov" should be flattened to ensure access to the referenced media.
Don't Flatten
Flatten

If you happen to upload a folder, a dialog appears asking how you'd like to upload the files inside the folder.

For example, if you are uploading a folder full of images that you would like to individually catalog, then you will want to create individual assets. However, if you have collected files into a folder for a specific purpose, then you may wish to create a single bundle asset for all of them.

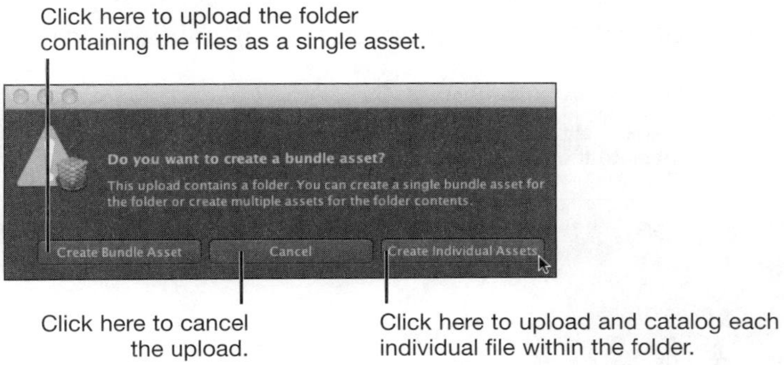

Adding Metadata During Upload

After you use either method to begin the process, an Upload window will appear, prompting you for several important pieces of information:

- The name of the copied file, usually kept as is.
- The device, and a path on that device, where you want the file to be placed, so that Final Cut Server can manage it.
- Whether the asset will be associated with a *production*—a virtual folder for organizing your assets by type of production (feature film, commercial spot, documentary sequence, or news package). More on productions in Lesson 4, "Working with Productions."

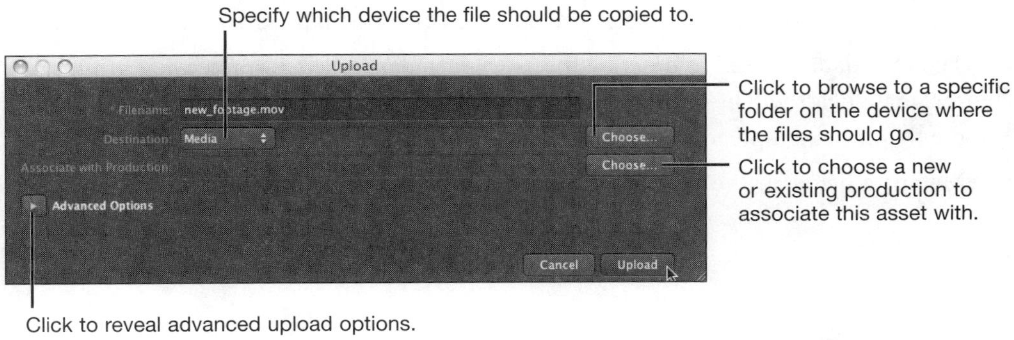

Specify which device the file should be copied to.
Upload
Filename
Destination
Media
Associate with Production
Choose...
Choose...
Advanced Options
Cancel
Click here to upload and catalog each individual file within the folder.
Upload
Click to browse to a specific folder on the device where the files should go.
Click to choose a new or existing production to associate this asset with.
Click to reveal advanced upload options.

Select a device from the Destination pop-up menu. It's important to know beforehand which device should receive your uploaded files, since every device varies in speed and capacity. There are also specialized devices, such as Library and Media, which might be appropriate. Be sure to talk to your administrator about the best device to store your uploaded files.

Once your device is selected, click Choose to specify the folder path for the file. An Open window appears. Since you are specifying a path on this device, you will only see folders on the device, not files.

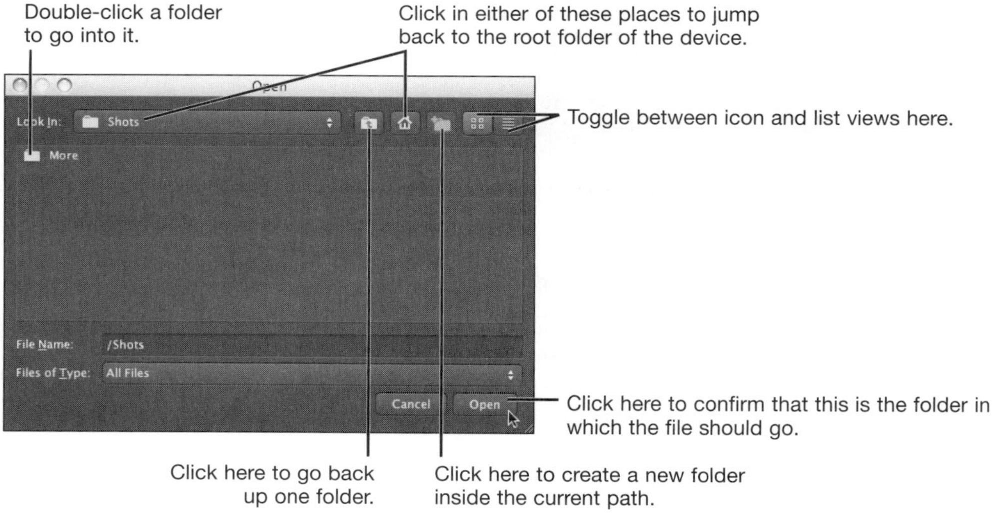
Double-click a folder to go into it.
Click in either of these places to jump back to the root folder of the device.
Toggle between icon and list views here.
Open
Look In:
Shots
More
File Name:
/Shots
Files of Type:
All Files
Cancel
Open
Click here to confirm that this is the folder in which the file should go.
Click here to go back up one folder.
Click here to create a new folder inside the current path.

Click the Advanced Options disclosure triangle to reveal additional options for uploading:

- A transcode setting, if needed, to convert the file into a different format when it uploads. Final Cut Server uses Compressor to do all of its video format conversions, and it has its own internal image processor to convert still images to different formats.
- The metadata set used to describe the asset. Metadata sets determine what groups of metadata fields will be used in the asset info window; they are based on the kind of file that is being uploaded.
- Metadata fields for the asset, which you can populate with custom information.

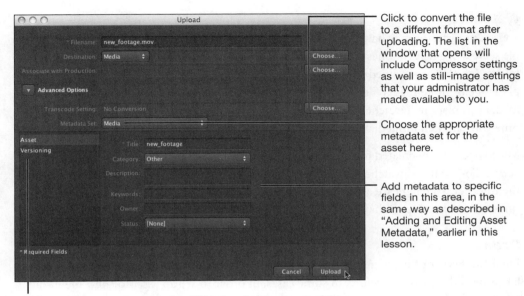

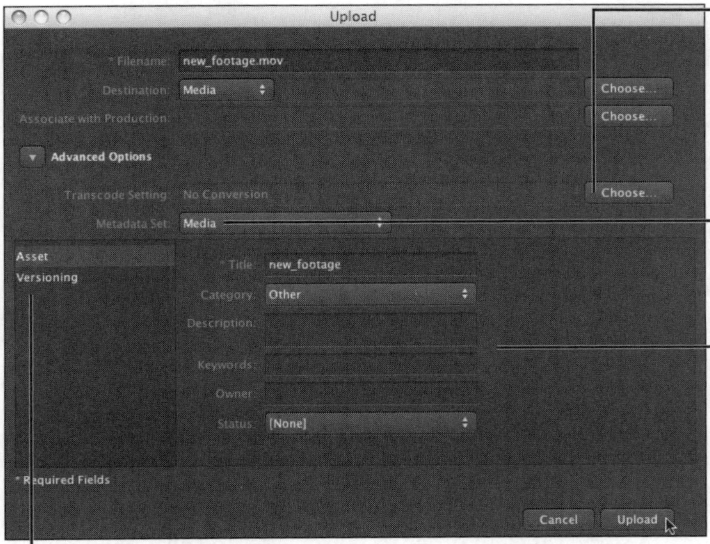

Finally, click Upload to start the process of uploading the file(s) to Final Cut Server. The progress of the upload can be viewed in the Job Status area of the client application window, located in the lower left.

Jobs in Progress
1
Job Status area
Click here to bring up the Downloads & Uploads window, which lists all your upload and download sessions since logging in.

NOTE ▸ If you drag multiple files into the client application's drop zone, you will get a slightly different window titled Multiple Upload, but its functionality is the same. All files that you drag in will upload using the settings you specify in the Multiple Upload window.

MORE INFO ▸ Final Cut Studio projects can, of course, be uploaded using these methods, and additional settings apply. These are explained in detail in Lesson 3, "Working with Final Cut Studio Projects."

Watchers

Your administrator may have created watchers, which are simply folders located on network file servers or Xsan volumes accessible by you. Final Cut Server constantly watches these folders for new or modified content. To upload files and create assets from a watcher, simply place one or more media files within them, and Final Cut Server will do the rest. Usually, the media you place in a watcher is deleted when Final Cut Server completes its tasks, at which time you can then immediately search and locate the new asset(s) in the client application.

Most watchers are created for the submission of media files by users not using Final Cut Server. For example, watchers could be created for folders on FTP sites that are exposed to the Internet. Watchers are also *active* processes, meaning that they will watch for files continuously and act upon that file as soon as they determine that it is complete.

Scans

Final Cut Server often creates assets automatically by *scanning* a storage device on a regular or periodic basis. Therefore, in some cases, in order to see an asset for new media, all you have to do is wait! Network file servers and Xsan volumes within your organization might have scan schedules already assigned by your administrator.

For example, if the Scratch Disk settings for Final Cut Pro are set to a folder on an Xsan volume, and if Final Cut Server scans that volume periodically, newly captured clips will eventually show up as assets in Final Cut Server automatically. Scans are therefore used on devices that have constantly changing content. They are also *passive* in that they only happen during their scheduled time, and not in between.

Getting Assets Out of Final Cut Server

Now that you know how to add, edit, and search for assets, it's time to learn how to get them out of the catalog and into your applications!

To work with an asset in its native application, two steps are involved:

- Make the primary representation of the asset available to your local system. You can do this either by being on a Mac connected to an Xsan volume or by putting a copy of the primary representation in a cache on your computer's hard drive.
- Open the asset in the desired application.

This section covers all the ways that assets can be brought out of Final Cut Server.

Previewing Assets

You can preview an asset by looking at its clip proxy if it's a video asset, or its poster frame if it's a still image.

There are two methods for viewing previews:

- Click an asset's identifying icon in the thumbnail view of a search result, or click the same icon in the asset's info window.

- Select View > Proxy (or View > Poster Frame for still images) from the asset shortcut menu.

MORE INFO ▸ On most computers, previewing a clip proxy will open the QuickTime Player installed on that machine.

Still images will usually open with the Preview application on the Mac and the Windows Picture Viewer on a Windows machine.

To listen to an audio file, select View > Original Media from the asset's shortcut menu.

Checking Primary Representation Availability

Final Cut Server indicates that you can use the drag-and-drop method to access an asset's primary representation (or its hi-res file) by making its icon brighter in a search results window. You have drag-and-drop access if the primary representation is stored on an edit-in-place device (like an Xsan volume) that you have mounted on your desktop, or if the primary representation is already stored in your local cache.

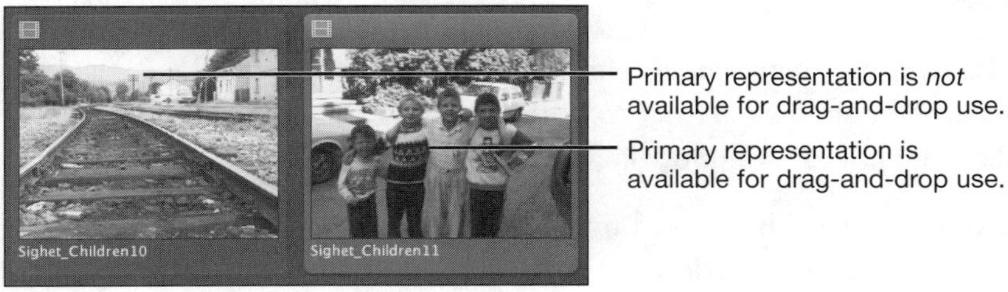
Sighet_Children10
Sighet_Children11
Primary representation is *not* available for drag-and-drop use.
Primary representation is available for drag-and-drop use.

In a list view of the search results window, the primary representation's availability is indicated by a light gray dot.

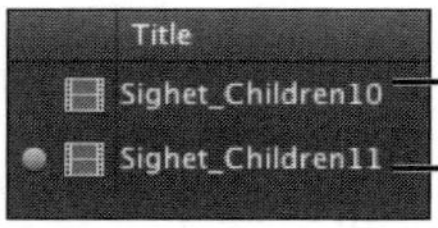

Primary representation is *not* available for drag-and-drop use.

Primary representation is available for drag-and-drop use.

The primary representation is usually available when you are running Final Cut Server on a Mac attached to an Xsan volume on which the primary representation resides. Your computer "sees" the file just as Final Cut Server does. This kind of file is called *edit-in-place*, since it doesn't need to travel anywhere to be used.

However, you can also make the primary representation available on a Mac that is not attached to an Xsan volume or a PC by simply adding the primary representation to your computer's cache.

NOTE ▸ We're using the term "adding to cache" to describe the process of downloading the asset's primary representation and placing it in a rather hidden place on your hard drive, called a cache. The actual file then stays in the cache, and when you drag and drop it using Final Cut Server's interface, a link for that file's location is passed to the location or application to which you drag the file.

Adding to Cache

To add an asset's primary representation to your cache, Control-click an asset or a selected group of assets and select Add to Cache from the asset shortcut menu.

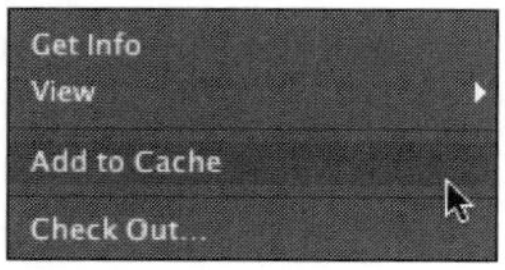

This will download the primary representation(s) automatically to your computer. Depending on your connection to Final Cut Server, this may take some time. The Job Status area of the main client application window will report the progress of the download.

When the download is complete, the asset icon will become a brighter shade of grey (as seen in the previous picture). (If you're looking at the asset in list view, you'll see a light gray dot appear to the left of the asset.)

Dragging and Dropping

Once an asset's primary representation is available, it can be dragged and dropped literally anywhere on your computer and will be available for use.

The following are some examples of places to which you can drag and drop an asset directly from a search result list:

- Your desktop
- Onto an application icon on your desktop or in a folder
- Onto an application icon in the Dock (Mac only)
- Into the Browser in Final Cut Pro
- Into the Asset tab in DVD Studio Pro
- Into the Project window of Motion

As you do so, you're simply creating a link between the cached or edit-in-place original representation and the place to which you drag and drop.

Just in case you forget to use Add to Cache before dragging and dropping an asset, Final Cut Server offers a friendly reminder. A handy Add to Cache button is offered to start the process immediately.

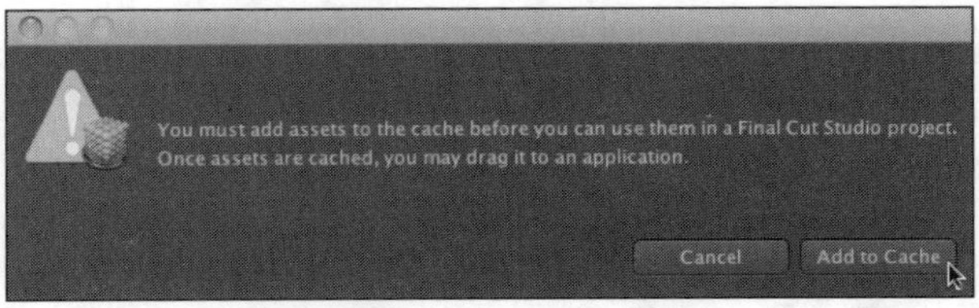
You must add assets to the cache before you can use them in a Final Cut Studio project.
Once assets are cached, you may drag it to an application.
Cancel
Add to Cache

Preparing for Disconnected Use

The Prepare for Disconnected Use feature allows Mac users to access the primary representations of assets specifically for adding into any Final Cut Studio project. This feature works the same way as Add to Cache, except that aliases of the cached files are conveniently placed in an easily accessible folder on your hard drive.

This way, they can be dragged and dropped to projects when you are away from your organization's network and therefore unable to run Final Cut Server. You still use the copy of the original representation that is located in the cache. The alias merely provides a convenient way to access the file.

To use this feature, Control-click an asset or a selected group of assets and select Prepare for Disconnected Use from the asset shortcut menu.

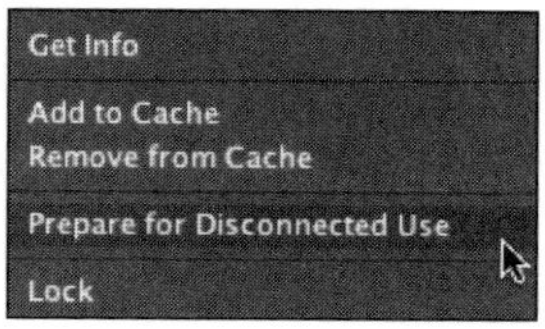

Once the process is complete, aliases for these files will appear in the following path within your home folder:

/Documents/Final Cut Server/Media Aliases

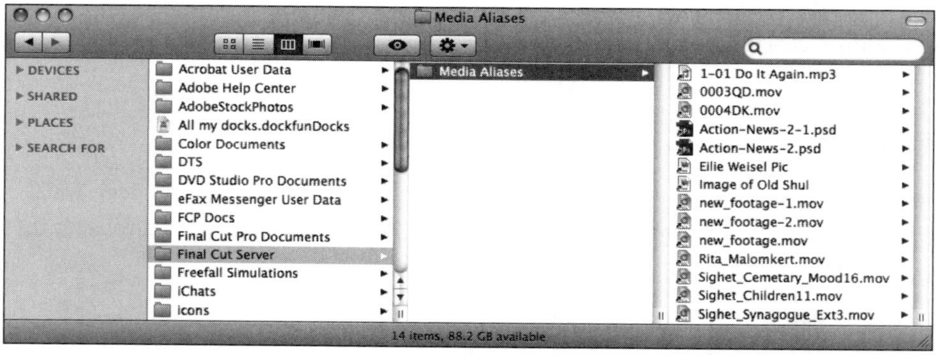

These aliases can now be used just like the original hi-res files, whether or not you are using Final Cut Server.

Adjusting the Cache and Alias Preferences

You can limit the size of your cache. The client application will automatically delete the oldest files from the cache in order to keep below the maximum size.

You can also change the location for the aliases made by the Prepare for Disconnected Use feature, and also clear the aliases that have collected in the current location.

To make any of these changes, select Preferences from the Server pull-down menu in the upper-left area of the main client application window.

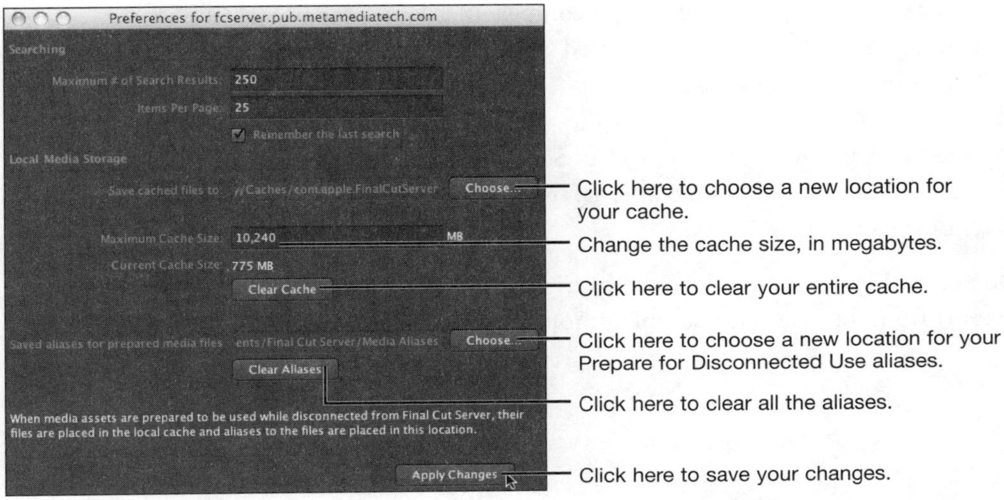
Preferences for fcserver.pub.metamediatech.com
Searching
Maximum # of Search Results: 250
Items Per Page: 25
Remember the last search
Local Media Storage
Save cached files to: ~/Caches/com.apple.FinalCutServer
Choose...
Maximum Cache Size: 10,240 MB
Current Cache Size: 775 MB
Clear Cache
Saved aliases for prepared media files: ents/Final Cut Server/Media Aliases
Choose...
Clear Aliases
When media assets are prepared to be used while disconnected from Final Cut Server, their files are placed in the local cache and aliases to the files are placed in this location.
Apply Changes
Click here to choose a new location for your cache.
Change the cache size, in megabytes.
Click here to clear your entire cache.
Click here to choose a new location for your Prepare for Disconnected Use aliases.
Click here to clear all the aliases.
Click here to save your changes.

Instead of clearing the entire cache, you can clear the cache for one asset or a group of assets by Control-clicking them and selecting Remove from Cache from the asset shortcut menu.

Get Info
View
Remove from Cache
Check Out...

Finally, you may launch Final Cut Server to discover that some of your previous cached assets have lost their cache files. An expired cache icon indicates this.

This can be fixed by selecting Remove from Cache and then Add to Cache from the asset shortcut menu.

Exporting Primary Representations

The Export command performs a no-frills download of the primary representation of an asset to a specific place on your computer. You also have the options of renaming it and transcoding it to another format before the download. Since exported files are copies, the original representation of the asset remains intact.

NOTE ► The exported file is also completely separate from Final Cut Server's catalog. If you make changes to an exported file, you will need to upload it to Final Cut Server in order to add it to the catalog.

As with the Upload function, Final Cut Server uses Compressor to do all video-format conversions, and it has its own internal image processor for converting still images to different formats.

To use the Export command, Control-click an asset or a selected group of assets and select Export from the asset shortcut menu. You are presented with an Export window.

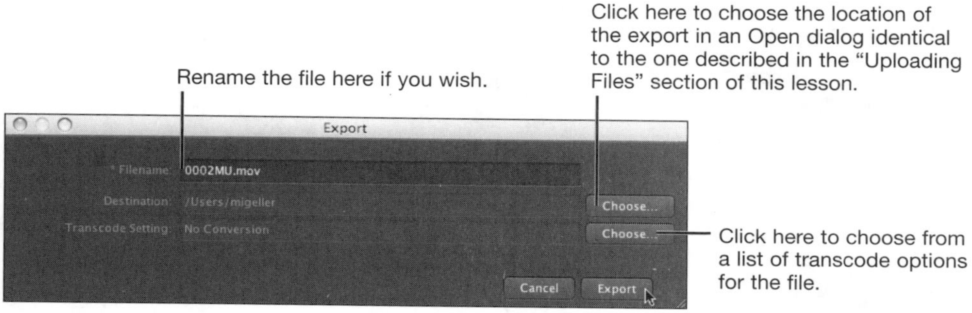
Click here to choose the location of the export in an Open dialog identical to the one described in the "Uploading Files" section of this lesson.
Rename the file here if you wish.
Export
Filename:
0002MU.mov
Destination:
/Users/migeller
Transcode Setting:
No Conversion
Choose...
Choose...
Click here to choose from a list of transcode options for the file.
Cancel
Export

There might be a slight pause if you elected to have Final Cut Server convert the primary representation to another format. As the representation downloads, the Job Status area of the main client application window reports the progress of the download. After the download is complete, you can open the file from its saved location.

NOTE ► If you've used Compressor before, you might notice that your transcode options are limited when exporting an asset. Contact your administrator, who has set these settings, if you wish to have more of them available to you.

Archiving and Restoring Assets

If your administrator has given you the ability to archive assets, the function will be available in the asset shortcut menu.

To use it, Control-click the asset or a selected group of assets and select Archive from the asset shortcut menu.

The primary representation of the asset(s) will be copied to the archive device at your facility. Afterward, the primary representation will be deleted from the original device on which it was located. The asset will continue to exist in the catalog, as will other representations such as clip proxies and thumbnails.

An archived asset has an identifying icon when shown in a search results list.

If a client or project requires use of an asset that has been archived, you can restore an asset or group of assets. There are two methods:

- Click on the archive icon of the asset in thumbnail view and select Restore from the resulting dialog.
- Control-click an archived asset or a selected group of archived assets and select Restore from the asset shortcut menu.

  This copies the primary representation back to its original location.

NOTE ▶ Some administrators may elect to store archived assets offline, on backup media such as tape, after they have been archived in Final Cut Server. In some cases, therefore, it may not be possible to restore the primary representation of an archived asset until your administrator, or an automated process, has restored it from backup. Talk to your administrator to find out the best way to restore such assets.

Duplicating Primary Representations

The Duplicate function does not copy assets to your computer. Instead, it is a request to upload the primary representations of an already cataloged asset from one *device* to another, creating a new asset in the process.

Therefore, you won't use the function very often, but it may come in handy for getting the primary representation of an asset to another device, especially since you have the ability to transcode it as it copies.

For example, if you want to quickly create a set of small clips on a file server from a series of large clips residing on an Xsan volume, the Duplicate function is for you.

You can even duplicate a primary representation back to the *same* device, provided that you rename it as you do. This would be helpful if you needed to create a copy of an asset in order to develop it in two different directions.

To use the Duplicate function, Control-click an asset or a selected group of assets and select Duplicate from the asset shortcut menu. You are presented with a Duplicate window.

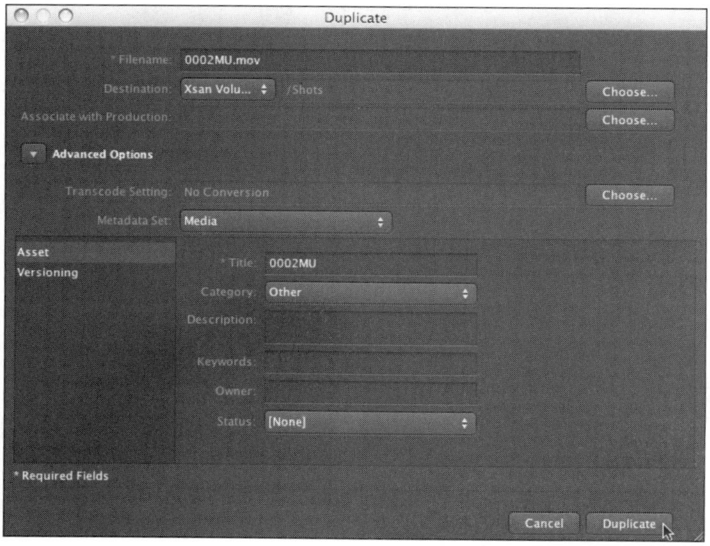

As you can see, the Duplicate window looks and functions identically to the Upload window, which we explored earlier in this lesson. A notable difference in technique, however, is that you may very well want to change the name of the asset as you duplicate it to another device. See the "Uploading Files" section for a description of the functions of this window.

Deleting Assets

Deleting assets removes not only the catalog entry, but also *all* representations of the asset. As such, deleting assets from Final Cut Server should be done with great caution, especially if they have not been archived first. Otherwise, the original file will need to be uploaded again.

There are two methods:

- Select the asset or group of assets and press Command-Delete (Mac) or Control-Delete (Windows).
- Select Delete from the asset shortcut menu.

In either case, a warning dialog confirms your action before the asset is deleted.

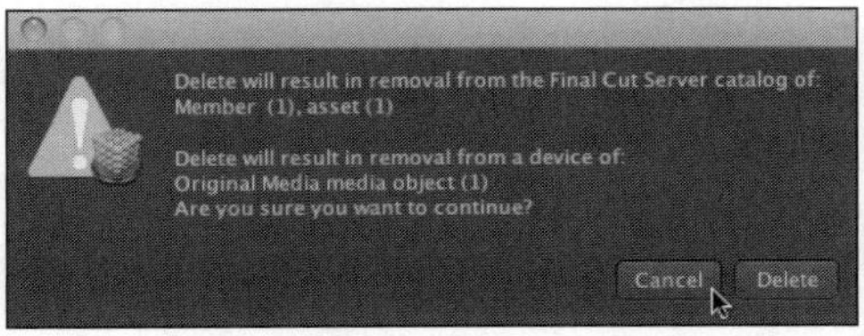
Delete will result in removal from the Final Cut Server catalog of:
Member (1), asset (1)
Delete will result in removal from a device of:
Original Media media object (1)
Are you sure you want to continue?
Cancel
Delete

Monitoring Access with Check Out and Version Control

When many users work on the same project or with the same assets, controlling access to a specific asset can help avoid confusion. Users can check out an asset to stop other users from making changes at the same time. Version control tracks reasons for changes and enables users to revert to a previous version.

Checking Out and Checking In Assets

Final Cut Server's *check out* feature is the primary working process with assets in Final Cut Server. It allows you to download the primary representation of an asset in order to make changes to it, while simultaneously locking the asset from being modified by another user in your organization. When the asset is ready to be checked in, Final Cut Server will automatically copy the primary representation back to the device it resides on, as long as the file did not change its location while you were working with it on your computer.

There are two methods you can use to check out an asset:

- Choose Check Out from the asset shortcut menu.
- Select the asset in a search results list and click the Check Out button in the Toolbar in the main client application window.

You will be presented with a Check Out window. Here you can choose the location on your computer's hard drive to place the original representation of the asset. You will be initially brought to your home folder. However, it's a good idea to designate a folder within your home folder to be the location for all of your checked-out assets.

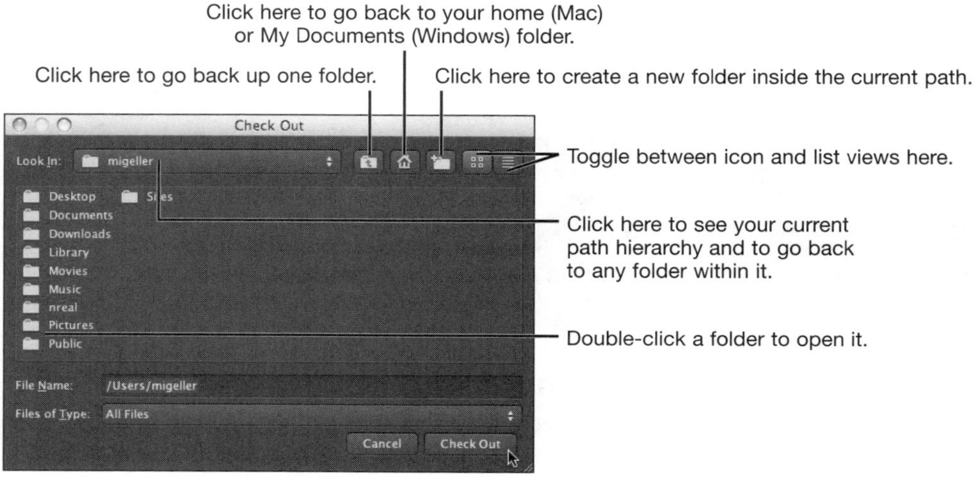
Click here to go back to your home (Mac) or My Documents (Windows) folder.
Click here to go back up one folder.
Click here to create a new folder inside the current path.
Toggle between icon and list views here.
Click here to see your current path hierarchy and to go back to any folder within it.
Double-click a folder to open it.
Check Out
Look In:
migeller
Desktop
Documents
Downloads
Library
Movies
Music
nreal
Pictures
Public
File Name:
/Users/migeller
Files of Type:
All Files
Cancel
Check Out

NOTE ► When checking out Final Cut Pro projects, the Save window will look slightly different. Please refer to Lesson 3, "Working with Final Cut Studio Projects," for specific details on checking out Final Cut Pro projects.

As the primary representation downloads, both the Job Status area of the main client application window and the icon itself report the progress of the download. After the download is complete, the asset will have an icon indicating that you checked it out.

You can now open the file from its saved location and begin editing it.

Checking in is even simpler. There are three methods:

- ► Click the checked-out icon directly on the asset's icon in thumbnail view.

- Choose Check In from the asset shortcut menu.
- Select the asset in a search results list and click the Check In button in the Toolbar in the main client application window.

The primary representation is automatically uploaded from its location on your computer and replaces the previous version of the original representation on the storage device. The asset is then reanalyzed from the new primary representation, replacing all other representations with new files. The Job Status area of the main client application window and the asset icon both report the progress of this process.

> **NOTE ▸** Remember to keep the asset's primary representation in the same place that you saved it when you checked it out. Check In works best if it can "pick up" the file from the place where it was last saved. If you move the file, you will get a dialog during check in asking for you to locate the file.

Sometimes you may accidentally check out an asset, or someone else at the facility may need to check out the same asset and it's clear that you haven't made any modifications to it. In these scenarios, if you want to cancel a checkout without having to check the asset back in (and replace the primary representation), select Cancel Check Out from the asset shortcut menu.

Revealing Checked Out Media

If you are unsure of the location of a checked-out asset's location on your hard drive, choose Reveal Checked Out Media from the asset shortcut menu. A Finder (Mac) or Windows Explorer (Windows) window will open to reveal the location of the checked-out file.

Tracking Changes with Version Control

If your administrator has enabled Final Cut Server's version control system, then it can be turned on for any asset. This allows you to save the previous version during a checkout/check-in cycle and add comments so others know what has changed in the new version.

MORE INFO ► Any asset that was uploaded using the client application will automatically have its version control enabled. However, assets that were originally added to the catalog by scanning or a watcher will need to have their version control enabled. Just make sure to turn it on by clicking the checkbox within the Versions pane, before you check in the asset.

Store and track versions of this asset

At check in, you will be prompted to provide notes about what has changed with the primary representation, so that others can understand the changes.

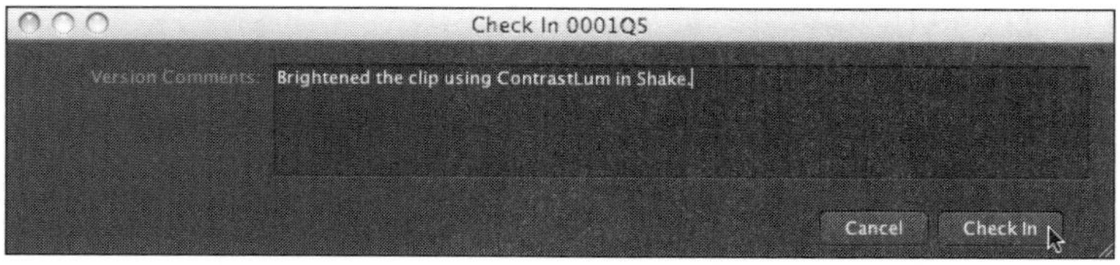
Check In 0001Q5
Version Comments:
Brightened the clip using ContrastLum in Shake.
Cancel
Check In

Once check-in is complete, the previous version will automatically be stored on a special device designated for previous versions of assets.

Other users can access information about previous versions, create new assets from the previous versions, and even revert to previous versions, within the Versions tab in the asset info window.

NOTE ► When you check in an asset, your version comments do not appear in the Versions pane until the next checkout/check-in cycle occurs.

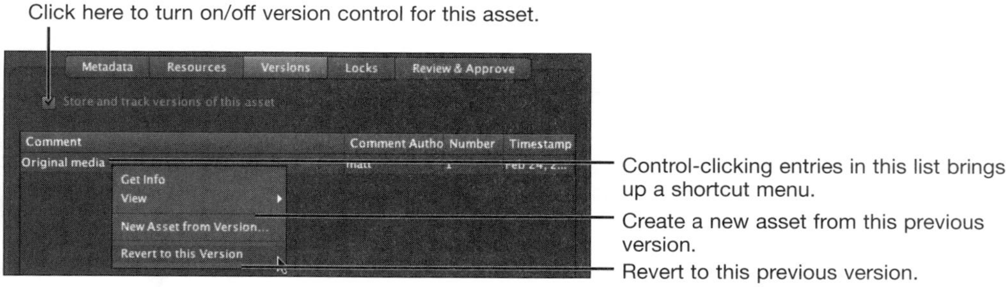
Click here to turn on/off version control for this asset.
Metadata
Resources
Versions
Locks
Review & Approve
Store and track versions of this asset
Comment
Comment Autho
Number
Timestamp
Original media
Get Info
View
New Asset from Version...
Revert to this Version
Control-clicking entries in this list brings up a shortcut menu.
Create a new asset from this previous version.
Revert to this previous version.

Lesson Review

1. Name two different ways that you can check out an asset.
2. When you upload a folder to Final Cut Server, what are your choices for bundling assets?
3. What is the difference between a watcher and Scanning?
4. True or false: Clip proxies are made for audio files.
5. What action do you take to download the primary representation of an asset to your computer?
6. What is generated from the primary representation when video files are uploaded to Final Cut Server?

Answers

1. Choose Check Out from the asset's shortcut menu, or select the asset and click the Check Out button in the toolbar of the main client application window.
2. In the "Do you want to create a bundle asset" dialog, you can choose to upload the bundle as a single asset, or upload and create individual assets for each of the files within that folder.
3. Watchers are folders on a device, like an Xsan volume, that process files immediately when they are placed inside. Scanning is the method that Final Cut Server uses to locate potential new assets, like those in a Final Cut Pro Scratch Disk folder. Scanning happens daily, weekly, or periodically.
4. False. Clip proxies are made only for video files. You can find previews for audio file assets by accessing the asset's shortcut menu and choosing View > Original Media.
5. Add to Cache. You can do this by choosing Add to Cache from the asset's shortcut menu. However, if the asset is on an Xsan Volume and your computer is an Xsan client, the asset's icon can be immediately dragged and dropped into any other application.
6. Poster frame, thumbnail, clip proxy (and an edit proxy if the administrator has enabled it).

3

Goals

- Upload Final Cut Studio projects
- Upload Final Cut Pro projects
- View Final Cut Pro project details and add metadata
- Check out and check in Final Cut Pro projects
- Export and duplicate Final Cut Pro projects

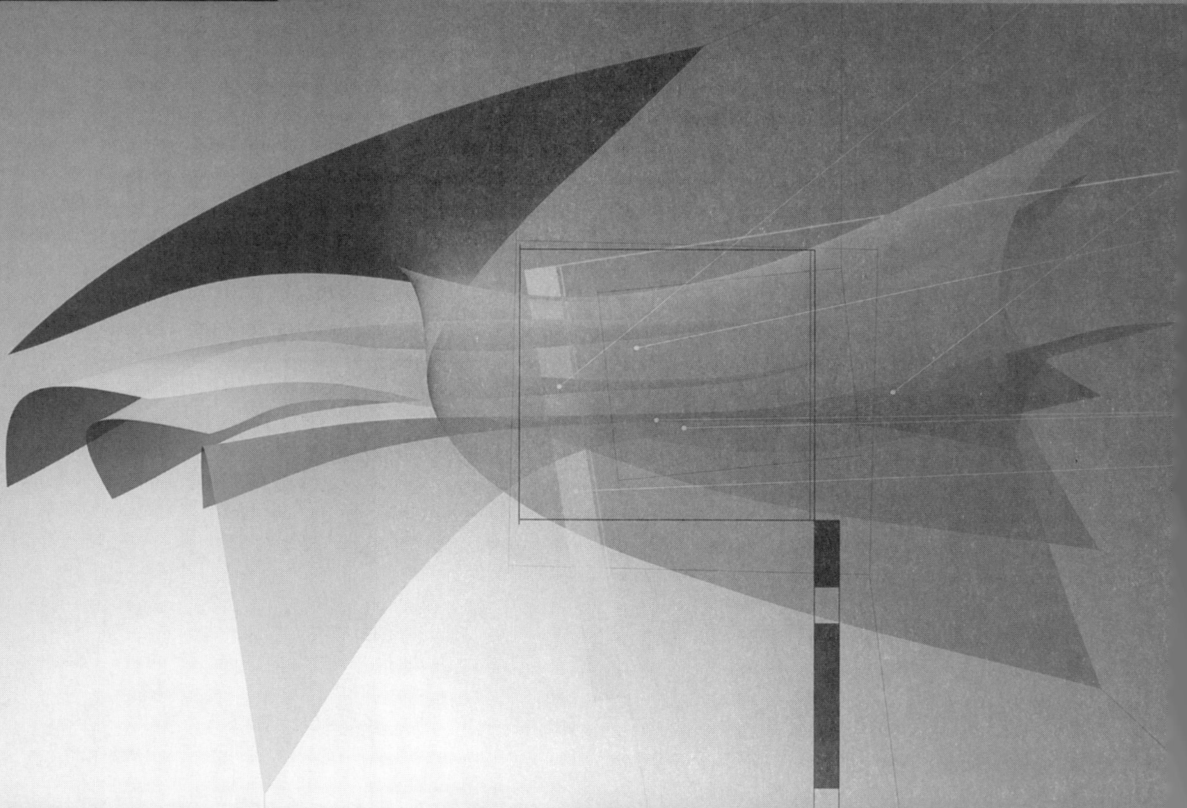

Lesson **3**

Working with Final Cut Studio Projects

The applications that make up Apple's Final Cut Studio—Final Cut Pro, DVD Studio Pro, Motion, Color, Soundtrack Pro, LiveType, Compressor, and Cinema Tools—each create unique project files.

Often times these projects are shared by many users within an organization, and this is where Final Cut Server can really help out. You can catalog these project files as you would other assets.

The benefits are immediate when Final Cut Server catalog becomes the storehouse for these project files, eliminating the tedious and often inaccurate filing systems that organizations have developed over the years for these kind of project files.

All of the features available to assets are also available to these project files. They can be locked, checked out, checked in, and exported, and they can be subject to version control.

Better yet, four specific types of project files within Final Cut Studio are intelligently recognized and are uniquely treated by the software.

Motion (.motn), DVD Studio Pro (.dspproj), and Soundtrack Pro Audio (.stap) project files are recognized during upload as being unique. This is because these kinds of project files are actually *bundles*, which may contain many other files within them. As a courtesy and as a means to manage the project file as a whole, Final Cut Server uploads the entire project file as a single bundle and then presents them as a single asset within the catalog.

Most importantly of all Final Cut Server will catalog a Final Cut Pro (.fcp) project file as a very special asset. The project file will be presented in the catalog as a container, holding additional assets within it. All of the clips, audio files, still images, and even sequences contained within the Final Cut Pro project file become assets as well.

The way Final Cut Server does this is to *parse,* or translate, the Final Cut Pro project file format into the XML (Extensible Markup Language) format. It then uses this parsed XML file to understand where all the linked media files are located and uploads those as well. This parsed XML file is saved as one of the representations of the Final Cut Pro project asset, and it later becomes an instrumental part of the collaborative process using Final Cut Server, as we will explore in this lesson.

When Final Cut Server uploads the video and audio clips and still image files contained within a Final Cut Pro project, it creates assets for each of them, just as it would if you uploaded them individually. The difference is that it then links them to and places them *inside* the Final Cut Pro project asset. Final Cut Server calls these linked assets *elements.* Once inside, the asset that was used to create the element is referred to as a *master asset.*

Adding Final Cut Studio Projects to Final Cut Server

Final Cut Studio application project files can be added as assets to Final Cut Server using the methods explained in Lesson 2, in the section "Getting Assets into Final Cut Server."

Uploading Project Files

Most project files upload as ordinary document assets. However, when uploading Motion, DVD Studio Pro, and Soundtrack Pro Audio projects, Final Cut Server recognizes them as bundles in order to manage them as a single asset. During the upload process, you will see a message reminding you that Final Cut Server will not automatically account for any media files that are not within that project file's bundle.

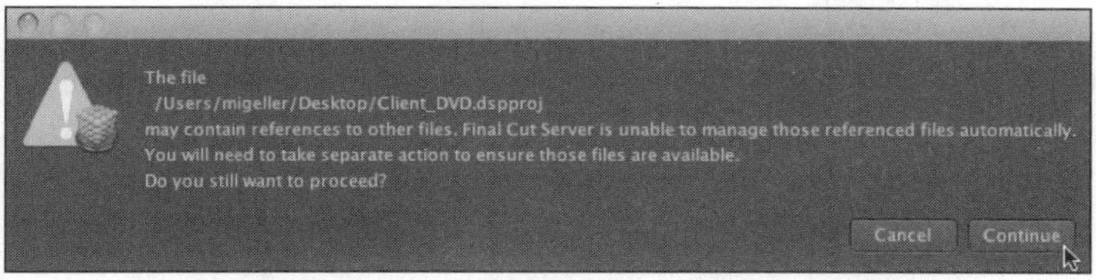
The file
/Users/migeller/Desktop/Client_DVD.dspproj
may contain references to other files. Final Cut Server is unable to manage those referenced files automatically.
You will need to take separate action to ensure those files are available.
Do you still want to proceed?
Cancel
Continue

This means that any media files associated with the project that reside locally on your computer should be uploaded in an additional step. If your organization often uses Final Cut Server to keep track of these kinds of projects, then creating productions for each of these projects will help. In productions, assets for project files and their associated media can be searched for and treated as a single entity. More on productions in the next lesson.

NOTE ▸ Make sure to use the Project metadata set when uploading project files. This metadata set contains fields specific to project files.

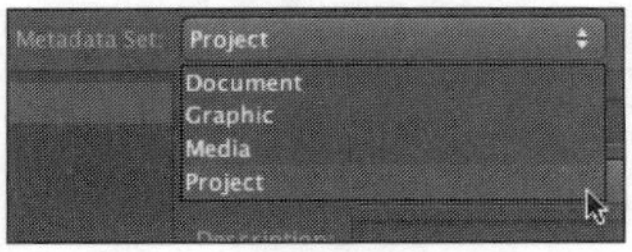
Metadata Set:
Project
Document
Graphic
Media
Project

Uploading Final Cut Pro Project Files

When uploading Final Cut Pro project files, Final Cut Server takes additional steps during the process:

- The Final Cut Pro project file is parsed into an XML file, which is then analyzed for all of the media files contained within it. All of these media files are linked to the Final Cut Pro project asset and become elements contained within it.
- If a discovered media file is already an asset in the Final Cut Server catalog, the element in the project is simply linked to it, and a new asset is not created.
- If no asset exists in the catalog for any element within the project file, those media files are automatically uploaded to the same device and path as the Final Cut Pro project file. Further, assets are automatically made for those media files, and links are established between the elements within the Final Cut Pro project asset and the new assets.

NOTE ▶ When getting Final Cut Pro project files into Final Cut Server, uploading will be preferred over placing the files in watchers or scanning, because uploading guarantees that assets not already part of the Final Cut Server catalog will get uploaded as well.

The regular Upload window is replaced with a specialized Upload Final Cut Pro Project window. At first glance, this window has the same features as the standard Upload window, and its functionality is the same as was described in Lesson 2.

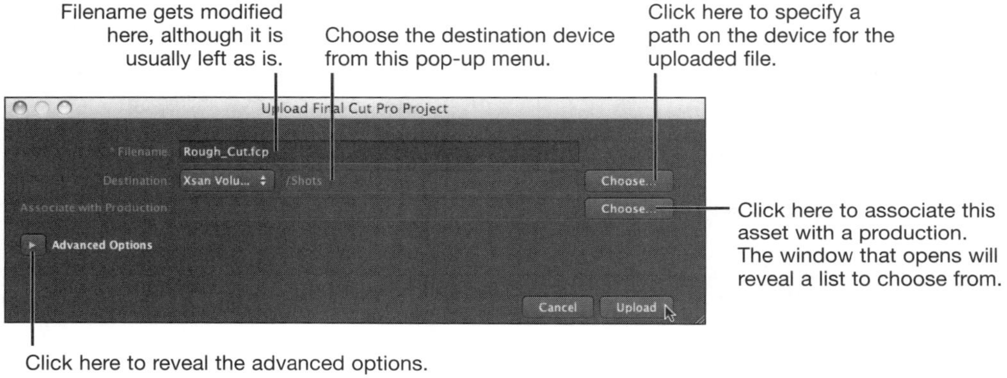
Filename gets modified here, although it is usually left as is.
Choose the destination device from this pop-up menu.
Click here to specify a path on the device for the uploaded file.
Upload Final Cut Pro Project
* Filename: Rough_Cut.fcp
Destination: Xsan Volu... /Shots
Choose...
Associate with Production:
Choose...
Click here to associate this asset with a production. The window that opens will reveal a list to choose from.
Advanced Options
Cancel
Upload
Click here to reveal the advanced options.

Clicking the Advanced Options button, at left, reveals a set of toggle buttons for two kinds of metadata: project and linked media.

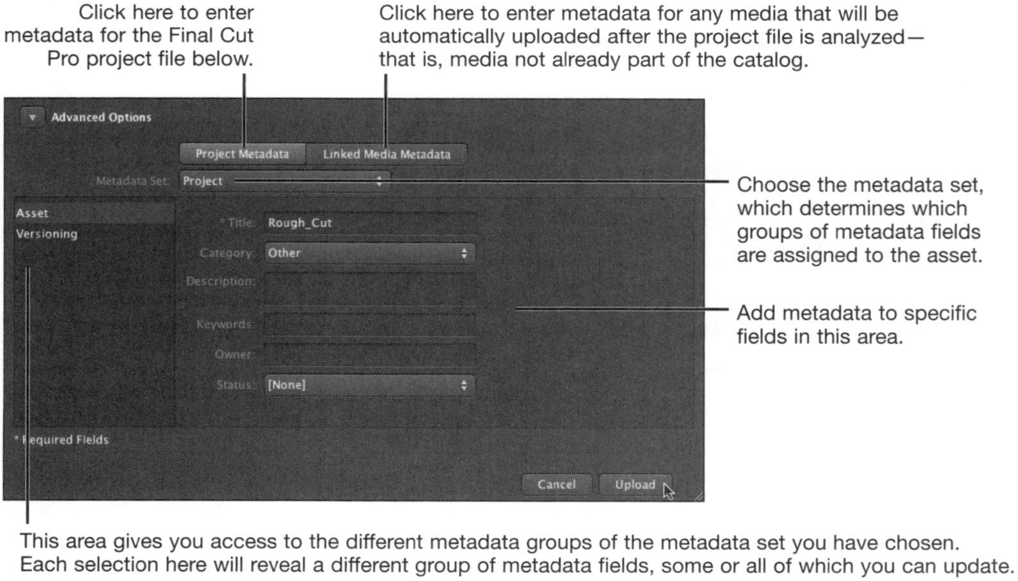
Click here to enter metadata for the Final Cut Pro project file below.
Click here to enter metadata for any media that will be automatically uploaded after the project file is analyzed—that is, media not already part of the catalog.
Advanced Options
Project Metadata
Linked Media Metadata
Metadata Set: Project
Choose the metadata set, which determines which groups of metadata fields are assigned to the asset.
Asset
Versioning
* Title: Rough_Cut
Category: Other
Description:
Keywords:
Owner:
Status: [None]
Add metadata to specific fields in this area.
* Required Fields
Cancel
Upload
This area gives you access to the different metadata groups of the metadata set you have chosen. Each selection here will reveal a different group of metadata fields, some or all of which you can update.

We fill out two sections of metadata for the two kinds of assets that will be made: the project file, and any associated media file (which Final Cut Server calls *linked media*) within the project asset. Remember that the associated media files become their own assets within the catalog and can be used for purposes other than being associated with this project. Because of this, taking the time to enrich the metadata fields of both the project asset and the associated media assets is of great benefit to the collaborative process. For example, simply entering keywords to unite all of the associated media allows for easy searching in the future.

> **NOTE ▸** You'll notice that the default metadata set for the Final Cut Pro project file is Project, and the default metadata set for the linked media files is, you guessed it, Media. Convenient!

Finally, click Upload to start the process of uploading the Final Cut Pro project, and its linked media, to the Final Cut Server catalog. The progress of the upload can be viewed in the Job Status area of the main client application window, located in the lower right.

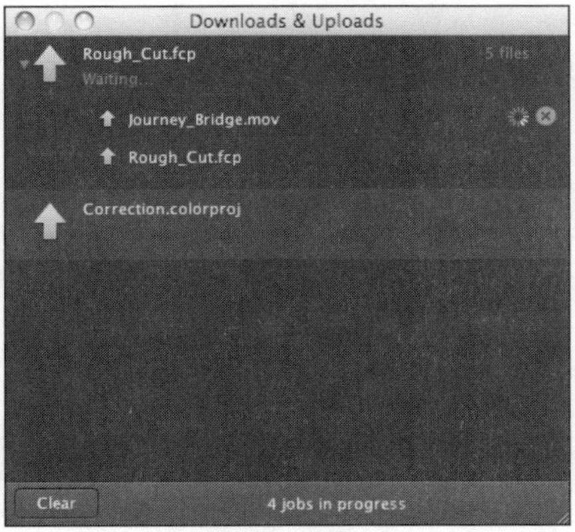

Understanding Version Control

All uploaded assets have version control automatically enabled. If you upload your Final Cut Pro project using the client application, you're ready to use version control. If not, make sure to enable version control by clicking the checkbox within the Version pane of the asset's info window.

When project files are checked out and then checked back in, the previous version of the file is saved in the version location, so that others can revert to that version or create a new asset from it. Comments can also be added during check-in to inform other about the changes in the new version. (See “Monitoring Access with Check Out and Version Control” in Lesson 2.)

Working with Final Cut Pro Project Assets

Now that we understand that Final Cut Pro project assets have unique status within the catalog, let’s explore the features that empower the collaborative process.

Viewing Final Cut Pro Project Asset Details

More information about the Final Cut Pro project asset, including a list of its elements, can be viewed inside its info window. There are two methods for viewing a Final Cut Pro project asset’s info window:

- Control-click (or right-click for Windows) the Final Cut Pro project asset and select Get Info from its shortcut menu.
- Double-click the Final Cut Pro project asset.

When the info window appears, you will see a new tab, called Elements, in the upper portion of the window. This new tab opens a pane that is like a miniature asset catalog, containing all of the elements within the Final Cut Pro project.

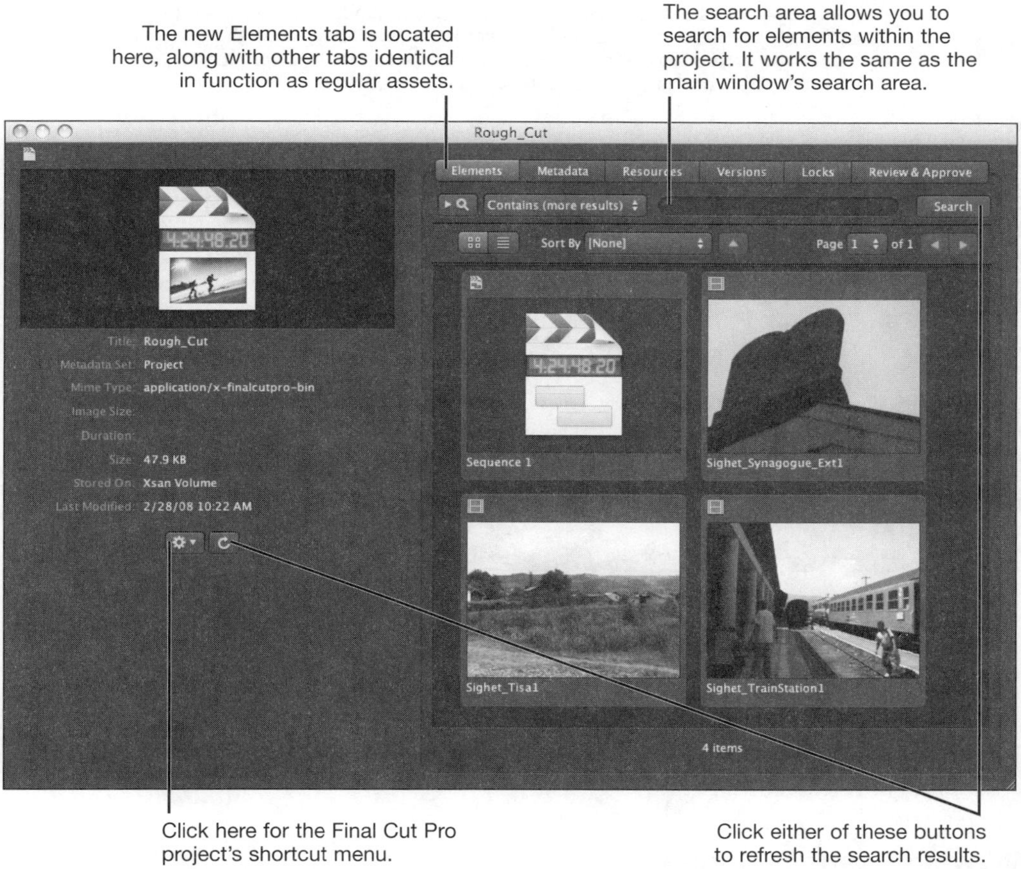
The new Elements tab is located here, along with other tabs identical in function as regular assets.
The search area allows you to search for elements within the project. It works the same as the main window's search area.
Rough_Cut
Elements
Metadata
Resources
Versions
Locks
Review & Approve
Contains (more results)
Search
Sort By [None]
Page 1 of 1
Title: Rough_Cut
Metadata Set: Project
Mime Type: application/x-finalcutpro-bin
Image Size:
Duration:
Size: 47.9 KB
Stored On: Xsan Volume
Last Modified: 2/28/08 10:22 AM
Sequence 1
Sighet_Synagogue_Ext1
Sighet_Tisa1
Sighet_TrainStation1
4 items
Click here for the Final Cut Pro project's shortcut menu.
Click either of these buttons to refresh the search results.

Adding and Editing Project Metadata

You can add and edit metadata in a Final Cut Pro project asset within the Metadata pane of its info window, following the same guidelines as for regular assets. (See "Adding and Editing Asset Metadata" in Lesson 2.)

Reviewing and Approving Final Cut Pro Project Assets

The Review & Approve tab has special significance for Final Cut Pro project assets because it can be used to track the status of the project, and it can hold a list of email addresses for people involved in the process of creating, editing, reviewing, and approving the content of the project. Depending on how your administrator has automated the function of the tab, it can be a powerful tool in the collaborative process.

Change the status of the project with this pop-up menu. Your administrator can modify the list of options you see here to more closely follow the process within your organization. He or she can also create automations that will trigger actions based on the status indicated here.

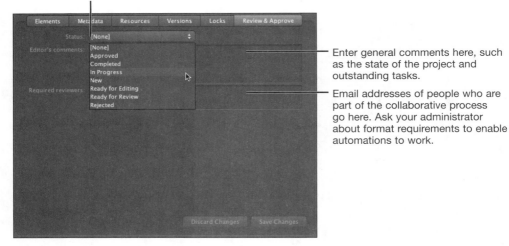

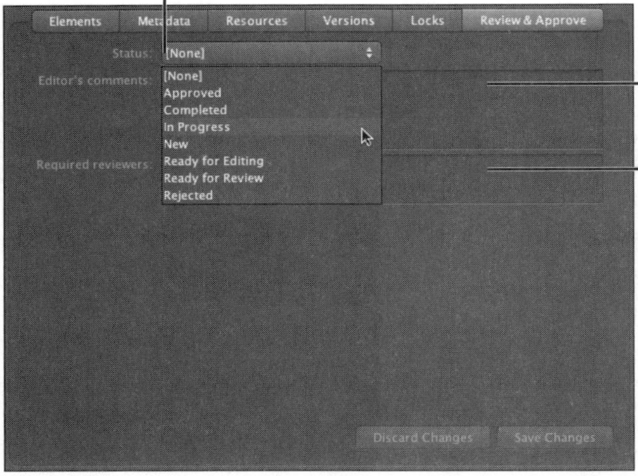

Enter general comments here, such as the state of the project and outstanding tasks.

Email addresses of people who are part of the collaborative process go here. Ask your administrator about format requirements to enable automations to work.

Working with Elements

Elements have a shortcut menu that you can bring up by Control-clicking an element in the Elements pane. Options within the shortcut menu are similar to those for assets, with a few exceptions explained below.

Viewing Element Details

Elements have a different info window than assets. There are two methods for viewing an element's info window:

- Control-click the element and select Get Info from its shortcut menu.
- Double-click the element.

The element info window has the same Metadata pane as its asset parent, but it also has a very informative Resources pane that shows its relationship to both the Final Cut Pro project that it is a part of and the master asset that it is related to. It also shows any other Final Cut Pro project asset or production in which the master asset is being used.

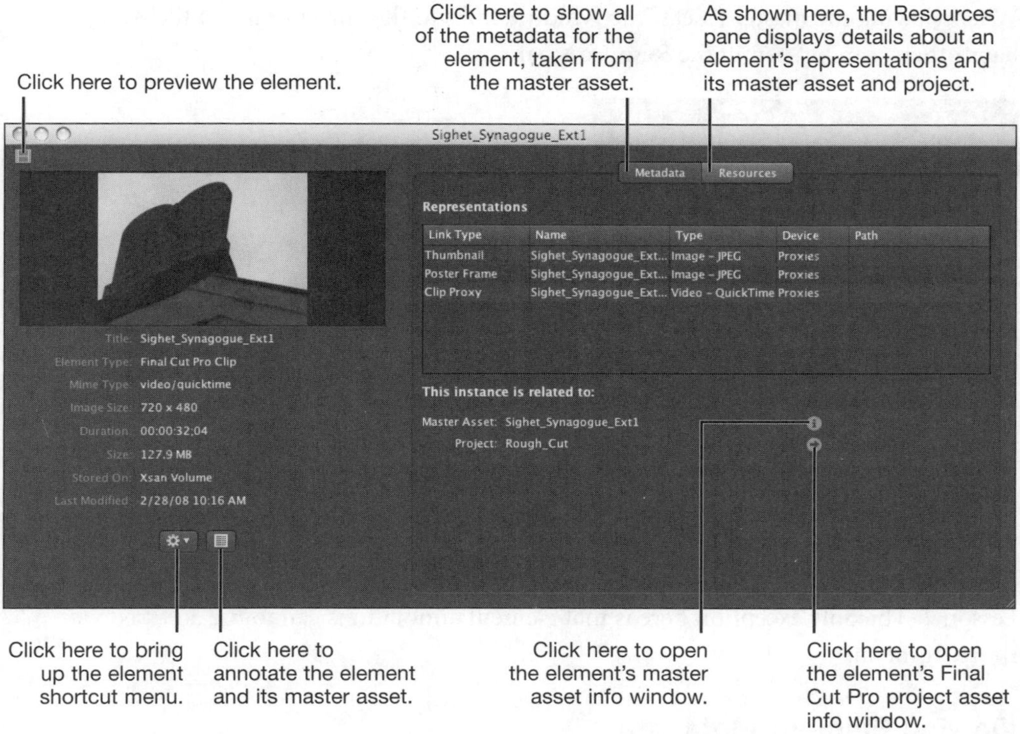

Viewing Master Assets

Viewing the master asset associated with an element can be done two ways:

- Control-click the element and select Show Master Asset from its shortcut menu.
- Click the master asset info button in the element's info window, found in the Resources pane.

When viewing the master asset's info window, links to the current project file, and perhaps others, can be seen in the Resources pane.

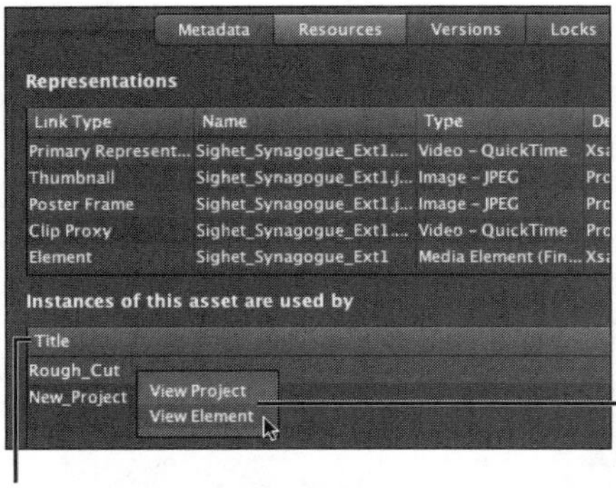

Control-click a project asset to show a shortcut menu, which can link you back to the project's or element's info view, respectively.

This area lists all Final Cut Pro project assets with which this asset is associated.

Annotating Elements

Elements are annotated in the exact same way as explained in "Annotating Clips" in Lesson 2. The only exception here is that element annotations cannot be accessed via shortcut menu.

Viewing Element Metadata

Elements within the Final Cut Pro project assets have no editable metadata fields. However, their Metadata pane contains a Final Cut Pro Logging metadata group, which displays all of the logging information for the element from the Final Cut Pro project file.

Checking Out and Checking In Final Cut Pro Project Assets

Final Cut Server's check out/check in feature is perfectly suited for Final Cut Pro project assets. Using this feature allows all members of the organization to share in the process of moving a project forward. Before we go into the details, let's review the entire process from stem to stern.

During checkout, Final Cut Server will automatically detect whether the computer you're on has real-time access to the media files associated with the project asset. As you'll recall from Lesson 2, *edit-in-place* devices, such as Xsan volumes, allow real-time access to their files for any Mac that is a client of the SAN. In this case, media files can just stay where they are. The project downloads to your computer's hard drive, and when opened, connects to the media files on the Xsan volume.

But users on non-SAN-attached Macs, such as laptops, usually don't have real-time access to edit-in-place devices such as Xsan volumes. In this case, Final Cut Server will automatically download the media files to your computer's cache, so that you can still work with the files. The only difference is that this takes time and space on your computer's hard drive.

The parsed XML version of the project file, created when the project file was originally uploaded and cataloged, is actually the representation used for the checkout process. Even though the checked-out file will have an .fcp extension at the end of its filename, its contents are XML. Final Cut Pro version 6.0.2 or later will recognize this file and open it up as if it were a regular project format.

Once you alter and then save the project file, its contents are returned once again to the standard Final Cut Pro project format.

During check-in, the standard .fcp project file is uploaded, reconverted to XML, and reanalyzed for any new media files. Any newly detected media files will automatically be uploaded, cataloged, and linked to the project asset.

If version control is enabled, checking in a project asset will also automatically save a copy of the previous version within the database. This ensures that you can revert to a previous version of the project for any reason. You can even "branch off" by making a previous version a new asset, with its own version control enabled.

Checking Out a Project

Final Cut Pro project assets are checked out as regular assets, as explained in "Monitoring Access with Check Out and Version Control" in Lesson 2. There are two methods you can use to check out an asset:

- Select Check Out from the project asset shortcut menu.
- Select the project asset in a search results list, and click the Check Out button in the Toolbar in the main client application window.

The Check Out window for Final Cut Pro project assets has a few unique settings.

Choose between original media (primary representations) or edit proxies here. If your administrator has enabled edit proxies, then your checked-out project file will connect to these files instead of the original representations. (Laptop and other non-SAN-attached users will download edit proxies instead of original representations.)

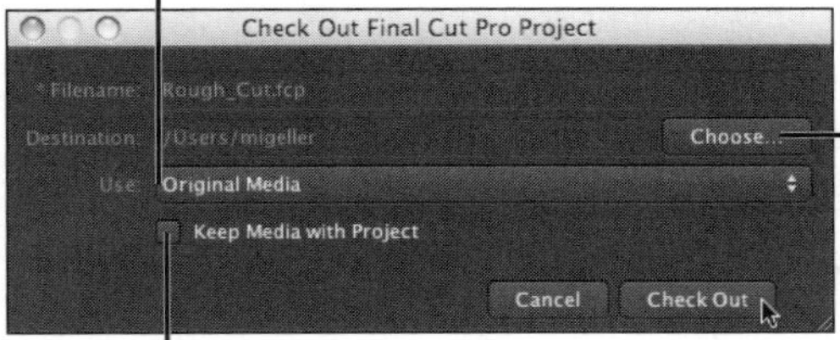

Click here to choose where to save the checked-out project file.

Click here to place both the project file and its associated media into a folder together. This checkbox is most suitable for non-SAN-attached Macs, such as laptops, and will place the checked-out project file and media files together in a handy folder. SAN-attached users should avoid this checkbox, since it will force Final Cut Server to download media that probably doesn't need to be.

MORE INFO ▸ For non-SAN-attached (laptop) users, to determine whether you should check out the Final Cut Pro project using the original media or the edit proxy, determine how much time and network bandwidth you have. If the media files are uncompressed, then the edit proxies may be an excellent alternative, especially when working on an Ethernet-attached machine. Remember that edit proxies are an optional feature in Final Cut Server, so they may not be available to you.

Click the Check Out button to save the project to the path you chose in the window. During checkout, media files will be managed in the manner that you selected:

- For laptop users and other non-SAN-attached Macs, media files download to your cache location by default, unless you chose Keep Media with Project, in which case they will download to the location you chose for saving the project file.
- For SAN-attached Macs, links will automatically be established to files on the Xsan volume, so no files need to download. Remember, however, that checking Keep Media with Project will force Final Cut Server to download the files to a location that you specify, so be careful when using this feature with SAN-attached Macs.

Finally, the Final Cut Pro project asset will receive its checked-out icon, notifying others who collaborate with you that you are working on the project.

Retrieving Original Final Cut Pro Projects

When you check out a Final Cut Pro project, you download the XML file representation. This file opens as if it were a standard project file in Final Cut Pro 6.0.2 or higher.

However, there are some rare circumstances where the XML version and the original standard project format will differ.

Not to worry. The original project file that was uploaded is always available as a manual download, just in case you need it:

1. Open the project asset's info window, either by double-clicking the project asset or selecting Get Info from its shortcut menu.
2. Click on the Resources tab at the top of the window.
3. Under the Representations section, find the Primary Representation of the project file. This is the original project file.
4. Right-click on this listing and select Retrieve Media. A Save dialog will then ask where to save the project file.

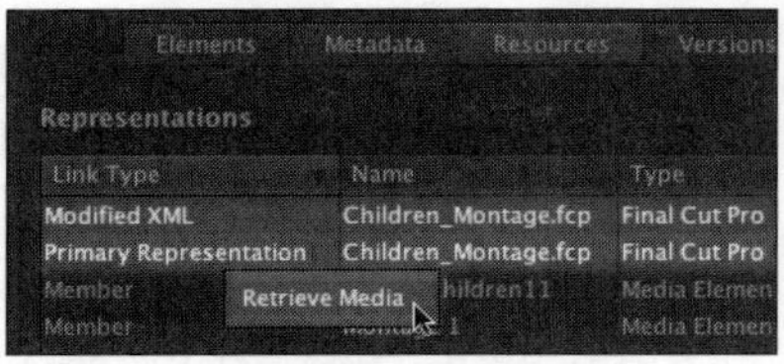

Checking In a Project

Any time you want to stop work on the project and update the Final Cut Server catalog with your changes, just save the project in Final Cut Pro. Then follow the steps below to check the project back in!

> **NOTE ►** Remember to keep the project in the same place as you saved it when you checked it out. Check In works best if it can "pick up" the file from the place where it was last saved. If you move the file, you will get a dialog during check-in asking for you to locate the file.

As with regular assets, there are three methods for checking in Final Cut Pro project assets:

- Click the checked-out icon directly on the Final Cut Pro project asset's icon.
- Select Check In from the asset's shortcut menu.
- Select the Final Cut Pro project asset in a search results list, and click the Check In button in the Toolbar in the main client application window.

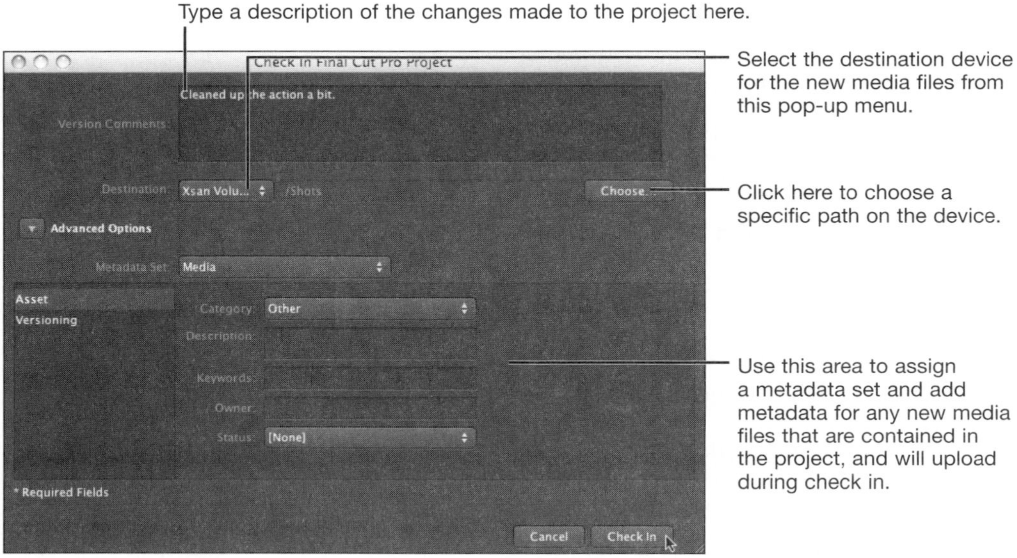
Type a description of the changes made to the project here.
Select the destination device for the new media files from this pop-up menu.
Click here to choose a specific path on the device.
Use this area to assign a metadata set and add metadata for any new media files that are contained in the project, and will upload during check in.
Check In Final Cut Pro Project
Cleaned up the action a bit.
Version Comments:
Destination:
Xsan Volu...
/Shots
Choose...
Advanced Options
Metadata Set:
Media
Asset
Versioning
Category:
Other
Description:
Keywords:
Owner:
Status:
[None]
* Required Fields
Cancel
Check In

After completing the information in this window, clicking the Check In button begins the upload process. The now-standard version of the project file is uploaded, converted to XML, and analyzed for new files. Any newly discovered files are also uploaded, made into assets, and linked to the project. Finally, the old version of the project is safely tucked away into the version control system.

When the check-in is complete, the Final Cut Pro project asset loses its checked-out badge, enabling it for checkout by the next member of the collaborative team.

Exporting Final Cut Pro Project Assets

When exporting Final Cut Pro project assets, the feature works exactly as it does for checkout, except that the asset is not marked with an icon or locked. What is handy about this feature is that a project and its associated media can be exported to your computer's hard drive for use offsite.

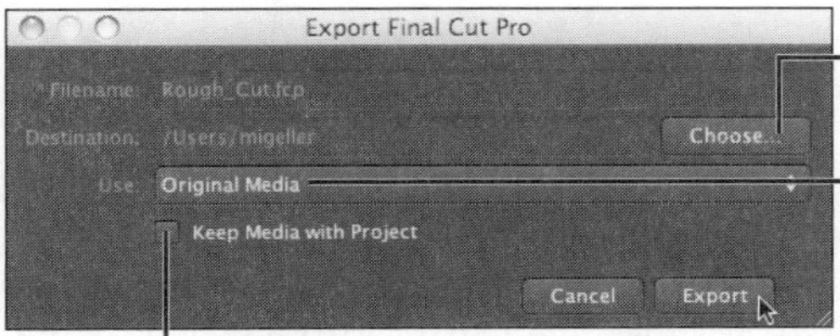

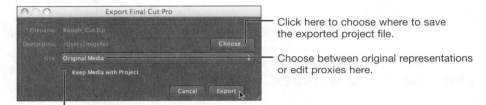

Click here to choose where to save the exported project file.

Choose between original representations or edit proxies here.

Click here to place both the project file and its associated media into a folder together. Remember that this checkbox should only be used if the media files are not accessible from edit-in-place devices, such as Xsan volumes, when requesting the export.

Using Media Prepared for Disconnected Use

As was explained in Lesson 2, the Prepare for Disconnected Use feature within an asset's shortcut menu allows you to cache the primary representations of media files locally to your hard drive, with the added benefit of creating easily accessible aliases.

Media files that are prepared for disconnected use show up as aliases in your Documents/Final Cut Server/Media Aliases folder. You can drag and drop these files into a Final Cut Pro project file as if they were the actual hi-res files.

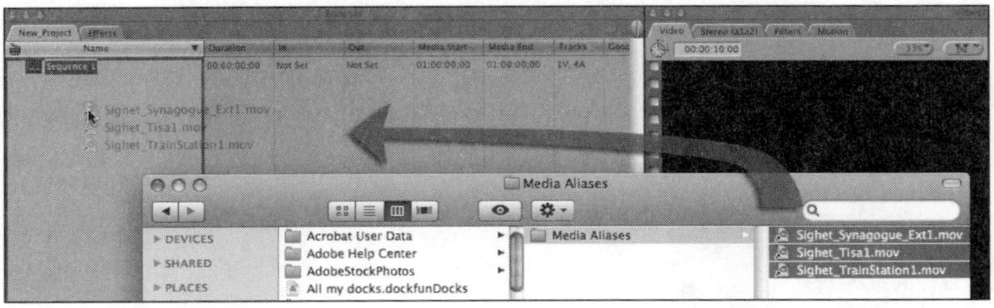

This feature works great for laptop-based users who may not always be within the walls of the organization, or even near the Internet to access the Final Cut Server application via VPN (for example, the park on a sunny day). After preparing files for disconnected use, you are free to use the files without using Final Cut Server.

More importantly, when you come back from the park and use Final Cut Server to check in a previously checked-out project file (or upload a new project file), you are guaranteed that these media files will not upload or create new assets, since they are already within the catalog. This is a huge time- and space-saving consideration.

NOTE ► See "Preparing for Disconnected Use" in Lesson 2 to find out how to prepare media for disconnected use.

Archiving Assets Associated with Final Cut Pro Projects

When you try to archive an asset that is associated with a Final Cut Pro project asset, Final Cut Server will warn you that you might be archiving an asset that is in use. Remember that the archiving process removes the primary representation of the asset from its original location on one of the organization's devices. Therefore, if someone wanted to work with the project mentioned in the dialog and needed to access an archived file, they would have to manually restore it before work with the file could continue.

However, if your task is to archive all of the assets that are associated with a particular Final Cut Pro asset, don't be daunted when this dialog shows the name of the Final Cut Pro project asset in question. You do need to be careful if the dialog mentions *more* than one project, since in effect you may be archiving assets that are in active use within other projects.

MORE INFO ► If the archive feature is dimmed in an asset's shortcut menu, it might be because the asset is being used within a Final Cut Pro project asset that is either locked or checked out. Look at the asset's Locks pane within its info window. If you see a "Reference" lock for the asset, it is protected from deletion or archive because it's being used by a project that is either locked or checked out.

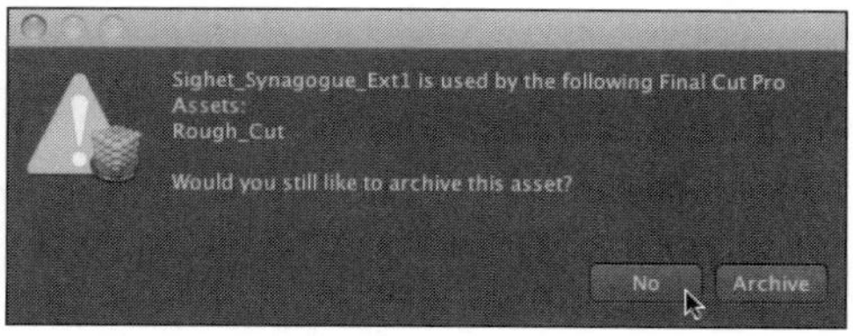
Sighet_Synagogue_Ext1 is used by the following Final Cut Pro Assets:
Rough_Cut
Would you still like to archive this asset?
No
Archive

Lesson Review

1. What's the difference between an element and an asset?
2. In the Final Cut Pro project upload window, what are the names of the two metadata tabs and what do they do?
3. True or false: Edit proxies are an optional feature in Final Cut Server.
4. What Final Cut Studio project files are recognized by Final Cut Server as ordinary documents rather than unique project files?
5. What will you find in the Resources pane of an element?

Answers

1. An asset is found in the Assets pane of Final Cut Server. Assets contained within a Final Cut Pro project asset are called *elements*. Elements, such as audio or video clips used in a project, are found within the Elements pane of the Final Cut Pro project asset info window.
2. Project Metadata and Linked Media Metadata. Project Metadata is where you enter metadata for the actual Final Cut Pro project file. Linked Media Metadata is where you enter metadata for the media that is uploaded after the Final Cut Pro project is analyzed.
3. True. Edit proxies don't have to be created for every video file linked to a Final Cut Pro project. If your administrator has not turned on the edit proxy feature, you will be able to check out only the original media, not the edit proxy.
4. LiveType, Cinema Tools, and Soundtrack Pro multitrack projects. If you upload these project files, the project will not include a list of elements for that project, as it will for a Final Cut Pro project.
5. The relationship to both the project that the element is part of and the master asset that the element is related to. Additionally, the pane tells you about each of its representations, such as its thumbnail, its poster frame, and its clip proxy.

4

Goals

- Search for productions
- Associate assets with productions
- Create productions
- View production details
- Add assets to productions
- Delete productions

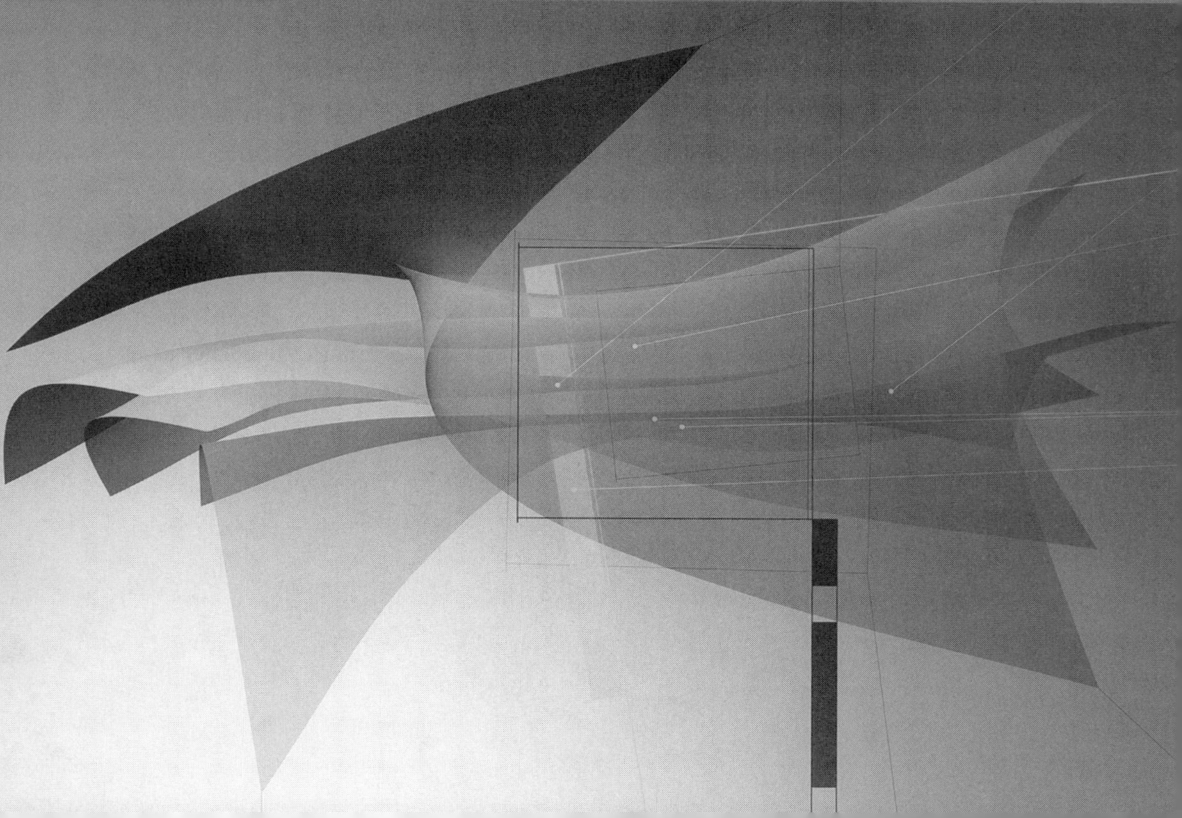

Lesson **4**

Working with Productions

A *production* is like a virtual folder in Final Cut Server: a simple yet powerful way to keep the assets for media, project files, and other documents organized within a single container. They can symbolize film projects, news packages, commercials, or an assemblage of commonly used "house" media.

Unlike folders in a traditional file system, production containers simply link to the assets they contain. So the same asset can be present in many containers without it being copied into multiple locations.

You'll probably first encounter a production as you associate uploaded files to one, but you can create one from scratch directly within the client application, adding assets into it by simple drag and drop.

As assets become associated with different productions, you will see these productions listed within the asset's info window.

Searching for Productions

You search for productions by clicking the Productions button in the upper left of the main client application window.

The search function works exactly as it does for assets, except the advanced search options are different.

With advanced searches, fields are used to filter searches in a similar fashion as with assets. For example, the Metadata Set field can be used to just return a particular kind of production.

> **NOTE ▸** If your administrator has customized the layout of the advanced search options, you may see more or fewer fields than are pictured here. If you require additional search fields in your facility, talk to your administrator about customizing the advanced search options.

Search results yield lists similar to asset searches.

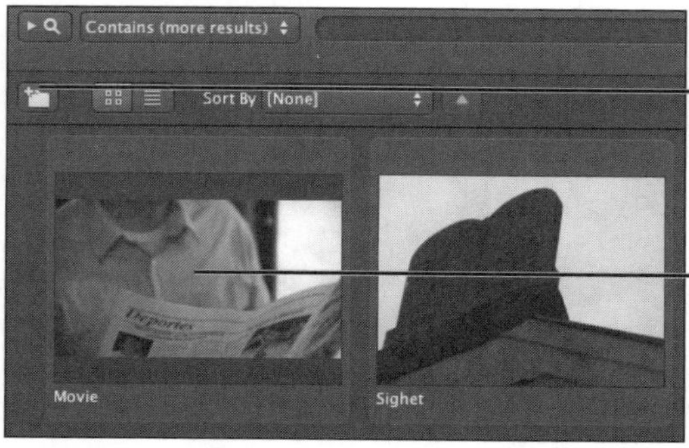

Click here to create a new production.

Productions in thumbail view appear as simple icons. They get their thumbnail from the first asset that is placed within them.

Associating Assets with Productions During Upload

You can use productions to gather together assets that already exist in your catalog, especially as you begin work. This will help you and your collaborators to focus on just the assets that are needed to get the job done.

Associating the assets of uploaded files to productions has been mentioned in Lessons 2 and 3. Here we'll look more closely at the functions of the specific buttons in the Upload window.

In the Upload window, click Choose, to the right of Associate with Production.

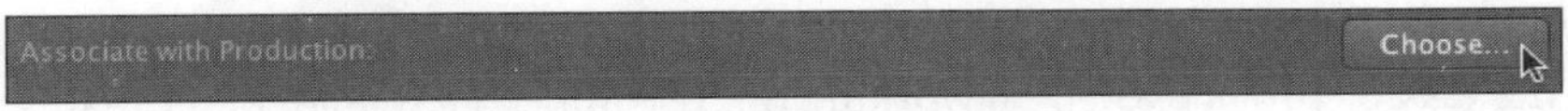
Associate with Production:
Choose...

A separate Find and Select window will open, displaying your existing productions. Find the production with which you wish to associate the asset(s) and click Open. This will make sure the newly uploaded asset(s) will be associated with the production in the catalog.

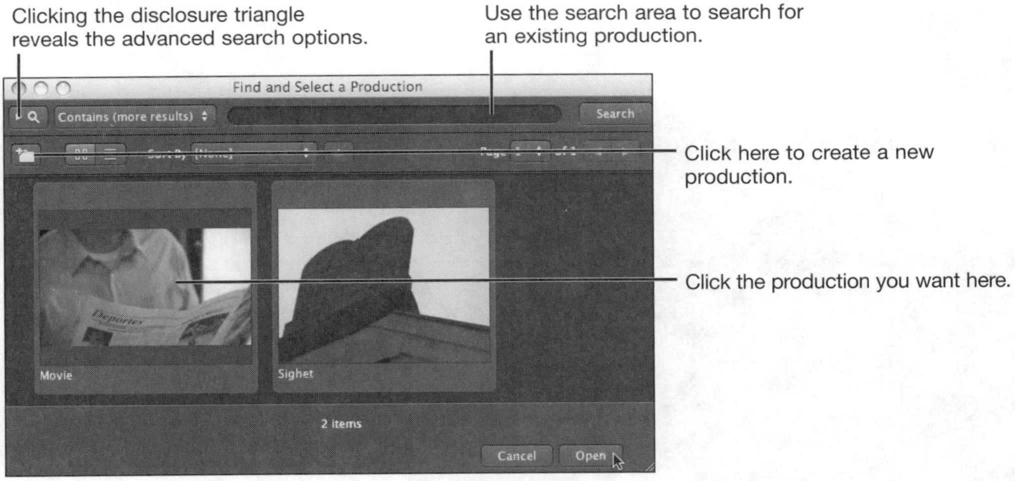
Clicking the disclosure triangle reveals the advanced search options.
Use the search area to search for an existing production.
Find and Select a Production
Contains (more results)
Search
Click here to create a new production.
Click the production you want here.
Movie
Sighet
2 items
Cancel
Open

If no production exists yet, you can click the Create button to make a new one, which we'll explain in the next section, "Creating Productions."

Click Open and your choice will appear back in the Upload window. If you made a mistake, there is a button provided to clear the field.

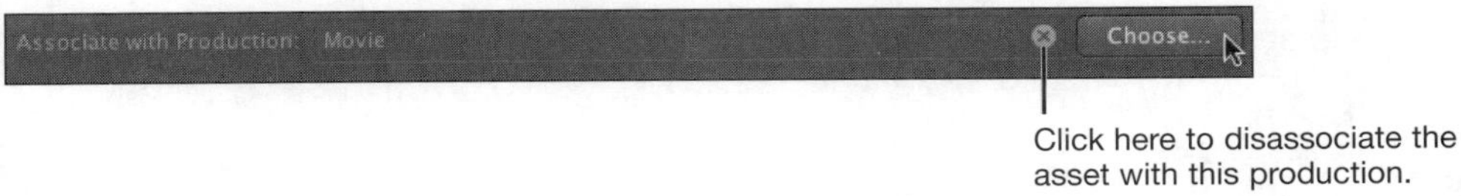
Associate with Production: Movie
Choose...
Click here to disassociate the asset with this production.

After upload, you will see all of your uploaded files' assets within the production's info window. (See the "Viewing Production Details" section, later in this lesson.)

Creating Productions

New productions are made with the Create button. This button is located either in the Production pane of the main client application window, or in the Find and Select a Production window during asset association.

When first creating a production, your first task will be to select a metadata set that is appropriate for the task or project you'll be working on. If you need a metadata set that is more appropriate, ask your administrator to create a custom one for you.

It's very important to add as much metadata here as you can before you save your changes, so that you and other collaborators can search for and find this production easily.

Click Create to open up the Production window.

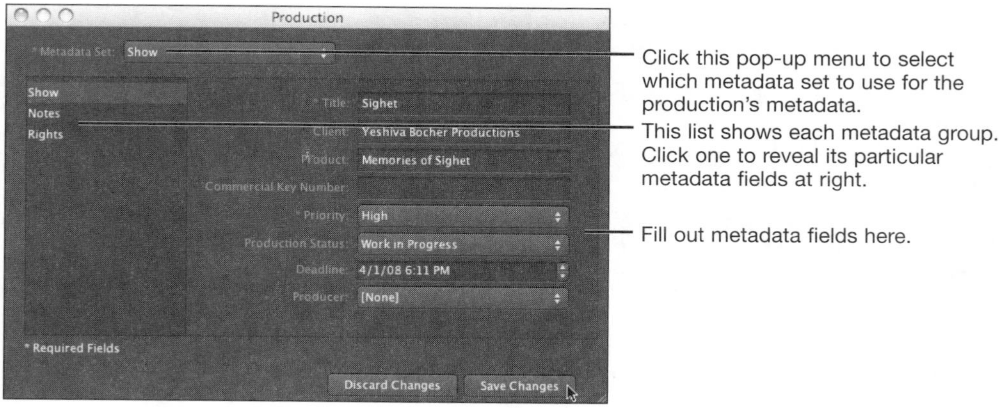
Production
Show
Notes
Rights
Sighet
Yeshiva Bocher Productions
Memories of Sighet
High
Work in Progress
4/1/08 6:11 PM
[None]
* Required Fields
Discard Changes
Save Changes
Click this pop-up menu to select which metadata set to use for the production's metadata.
This list shows each metadata group. Click one to reveal its particular metadata fields at right.
Fill out metadata fields here.

After clicking Save Changes, perform a search for your newly created production.

Viewing Production Details

There are two methods for opening the production info window:

- Double-click a production in a search results window.
- Select Get Info from the production's shortcut menu.

The production info window appears, which has two panes. The Assets pane displays all of the assets contained within the production. Besides its search area, it also has a toolbar for locking, checking out, and gathering assets to form other productions. The Metadata pane displays all of the metadata related to the production.

Toolbar icons here work the same as they do in the asset pane of the main application window.

Click here to see the assets contained within the production.

Click here to see the production's metadata.

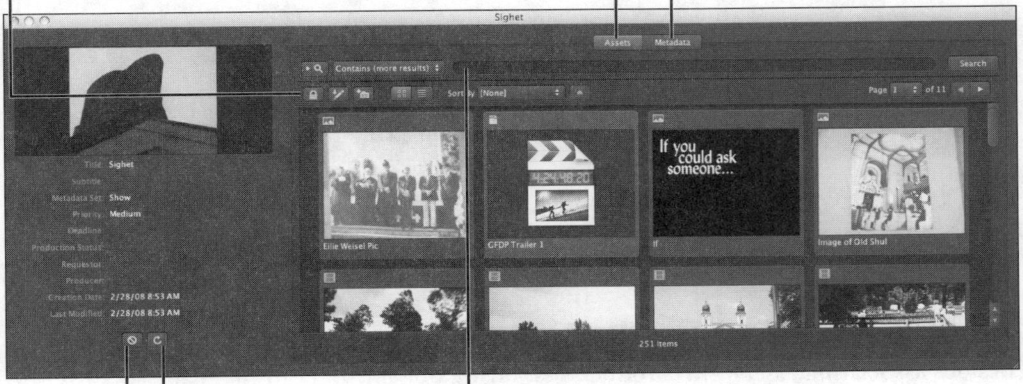

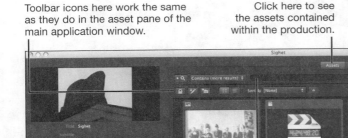

Click here to delete the production.

Click here to refresh the info window.

Use this search area to search for assets within the production.

Assets contained within the production can be used just like assets within Final Cut Server's main client application window. Their info windows have links to the production, found in the Resources pane.

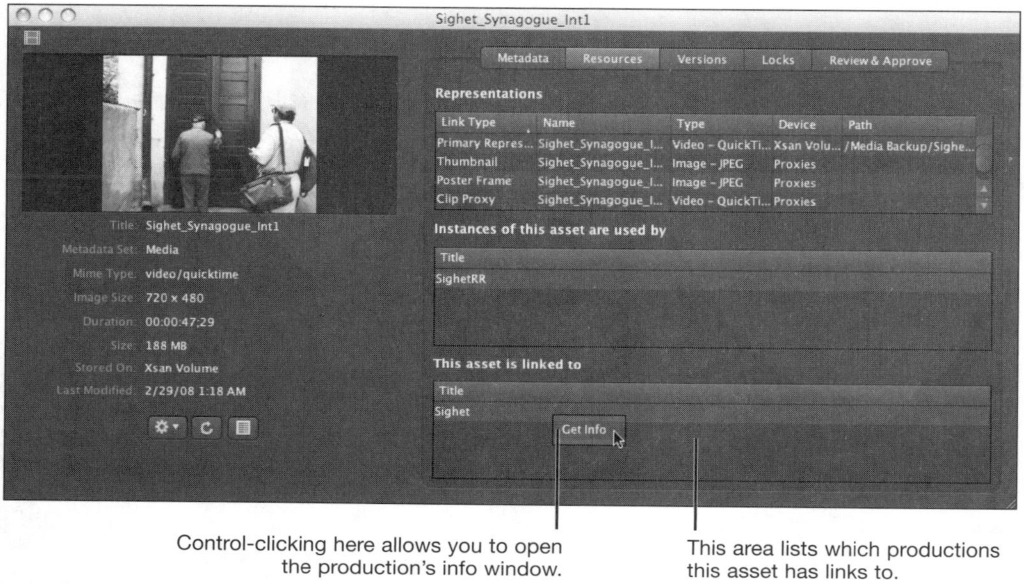

Adding Assets to Productions

Besides associating assets with productions during upload, you can manually add assets to productions by simply dragging them from a search results window and dropping them into the Assets pane of a specific production info window.

To make dragging and dropping easier, you can create a separate window by selecting New Workspace from the Window pull-down menu in the upper-left corner of the main application window.

With two windows open, you can set one to the asset pane and the other to production pane. Using this method, you can drag a series of assets directly onto the icon for a production. The icon becomes highlighted in blue to indicate that the assets will be added to that production.

If you wish to upload, catalog, and associate new files directly into a production, you can do so by simply dragging and dropping the files directly from the Finder, the desktop, or from a connected volume onto a production's icon or into its asset pane in the info window.

You can also Control-click an asset and choose New Production from Selection from the asset shortcut menu.

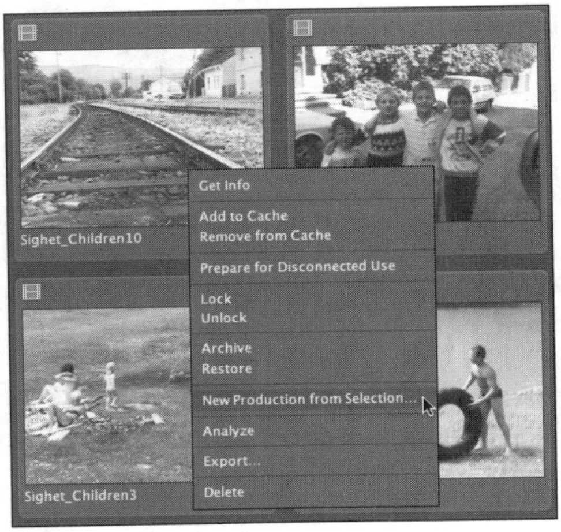

Finally, you can select a series of assets and click the New Production from Selection button in the Toolbar.

In these last two examples, you will be taken to a New Production window, where you can create the production to which these assets will be associated.

Deleting Productions

Deleting productions will *not* delete the assets contained within them, thankfully. Obviously, once a production is deleted, all assets disassociate from it and go their separate ways, like the breaking up of a rock band.

There are three methods for deleting a production:

- Select the production in a search results window, and press Command-Delete.
- Select Delete from the production's shortcut menu.
- Click the Delete button in the production's info window.

You will be prompted to confirm the deletion.

Delete Sighet?
This action cannot be undone.
Cancel
Delete

Lesson Review

1. Are assets always associated with productions?
2. True or false: Deleting productions will delete the assets that are contained inside the production from the Final Cut Server catalog.
3. What does the asterisk signify within a metadata window?
4. How do you create a New Workspace in Final Cut Server?
5. How can you find out what production an asset is linked to?

Answers

1. No. You can associate an asset with a production by choosing Associate with Production in the upload window, or you can drag and drop an existing asset from the search results window into a production.
2. False. If the production contains assets upon deletion, none of the assets are deleted from the catalog.
3. The field must be filled out with metadata. There are always some mandatory fields a user must fill out, like the type of metadata set or the title, but an administrator can make other fields mandatory if he or she wishes.
4. You can select New Workspace from the window pull-down menu in the upper left corner of the main application window. This is helpful if the user wishes to have the Assets pane and Productions pane open simultaneously.
5. By opening the asset's info window and clicking the Resources tab. A list of productions is displayed at the bottom.

5

Goals

- Search devices
- View device file details
- Catalog a device file
- Exporting and duplicating device files
- Delete files

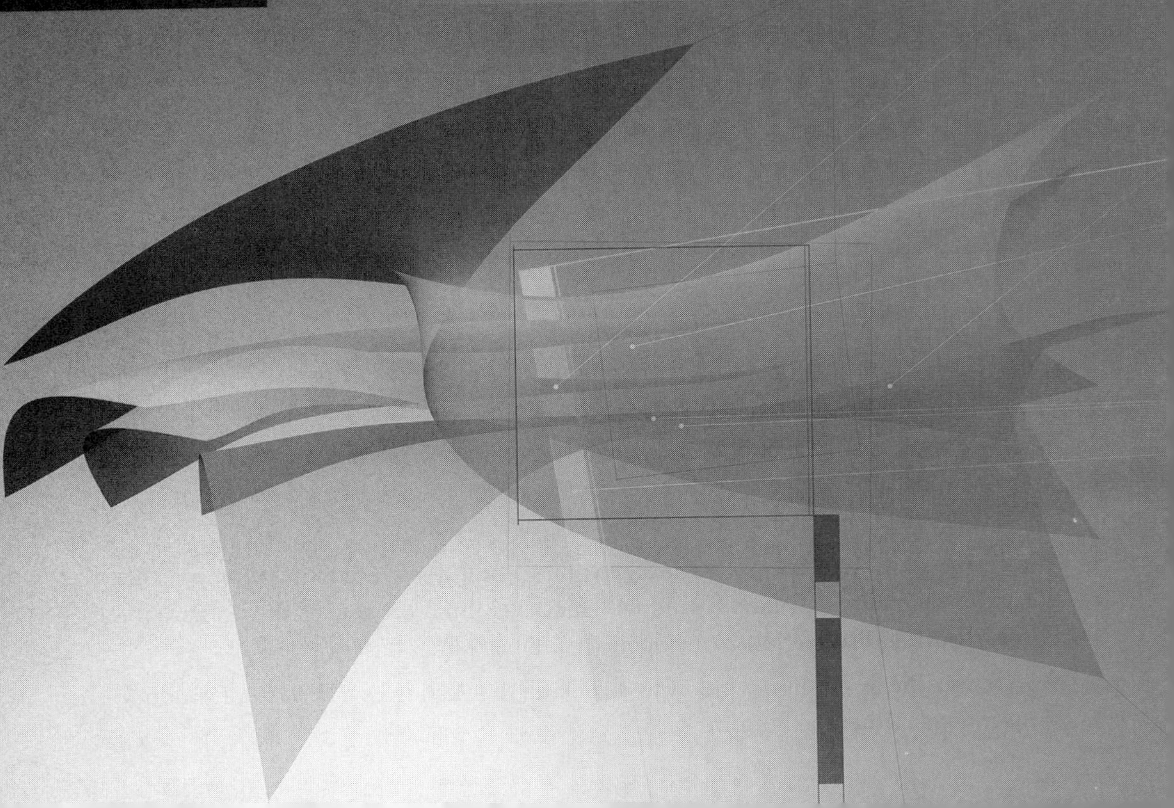

Lesson **5**

Working with Devices

Final Cut Server's catalog is designed to contain assets for every file that your facility uses on a regular basis. The devices, or storage locations, that Final Cut Server has been configured to recognize, may sometimes hold media files that are not yet an asset in the catalog.

This can happen when a file is brand new, or because Final Cut Server is not regularly scanning the device that the file is on (for example, a playout server or an FTP server that is exposed to the Internet).

Fortunately, Final Cut Server makes it easy to search for, download, copy, and manually catalog the files on these devices, provided that your administrator has given you access to do so.

Searching a Device for Files

You can search any device recognized by Final Cut Pro by selecting Search Devices from the Server pull-down menu located in the upper-left of the client application window.

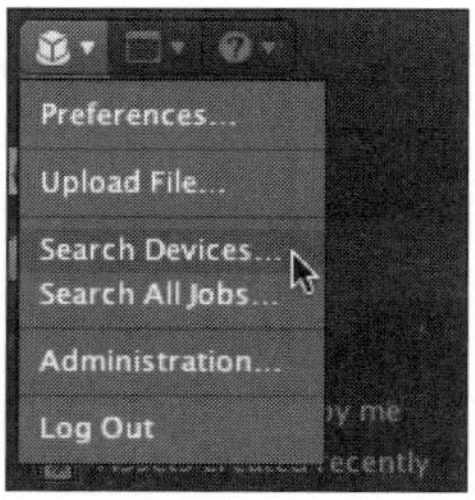
Preferences...
Upload File...
Search Devices...
Search All Jobs...
Administration...
Log Out

The Search Devices window appears.

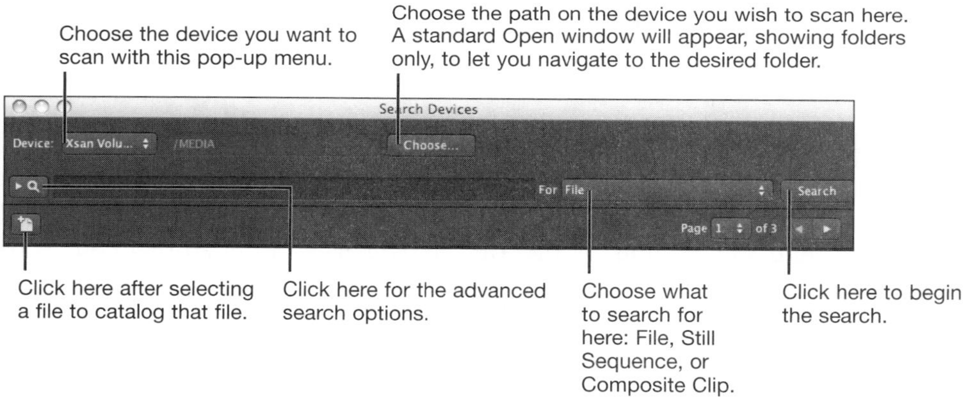
Choose the device you want to scan with this pop-up menu.
Choose the path on the device you wish to scan here. A standard Open window will appear, showing folders only, to let you navigate to the desired folder.
Search Devices
Device:
Xsan Volu...
/MEDIA
Choose...
For
File
Search
Page
1
of 3
Click here after selecting a file to catalog that file.
Click here for the advanced search options.
Choose what to search for here: File, Still Sequence, or Composite Clip.
Click here to begin the search.

The For pop-up menu has three items:

- *File*—Search for ordinary, single files.
- *Still Sequence*—Search for an image sequence, which has some sort of iterative numbering system in the filename. The files will appear as a single entry in the search results.
- *Composite Clip*—Search for an RGB clip and its corresponding, separate alpha channel clip. Both files will appear as a single entry when they return a result.

NOTE ▸ Although the Search Devices window assists you in searching for still sequences and composite clips, you can only catalog items found while searching for files.

Clicking the disclosure triangle at left will reveal the advanced search options, which allow a further refined search.

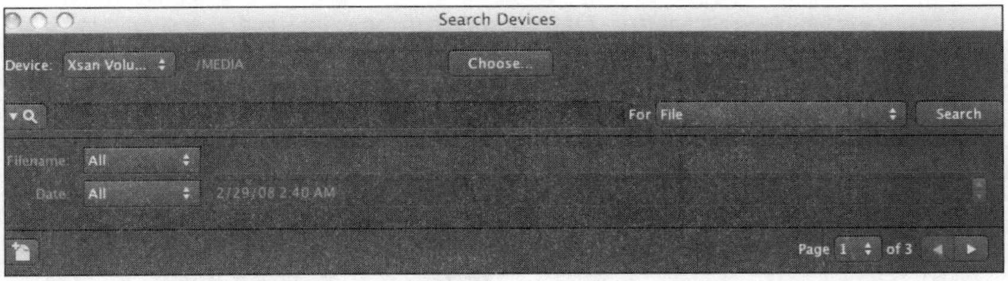

NOTE ▸ If your administrator has customized the layout of the advanced search options, you may see more or fewer fields than are pictured here. If you require additional search fields in your facility, talk to your administrator about customizing the advanced search options.

Viewing Device Content

Search results from the device and folder you have chosen will appear as a list in the area below. Sorry, no pretty thumbnail or tile views here. What do you expect with uncataloged files?

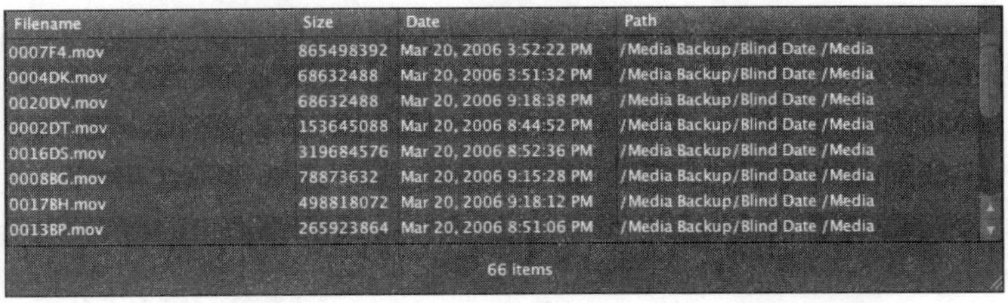

Viewing File Details

There are two methods to view additional file details:

- Double-click the file in the search results list.
- Control-click the file and select Get Info from the file's shortcut menu.

The file's info window appears.

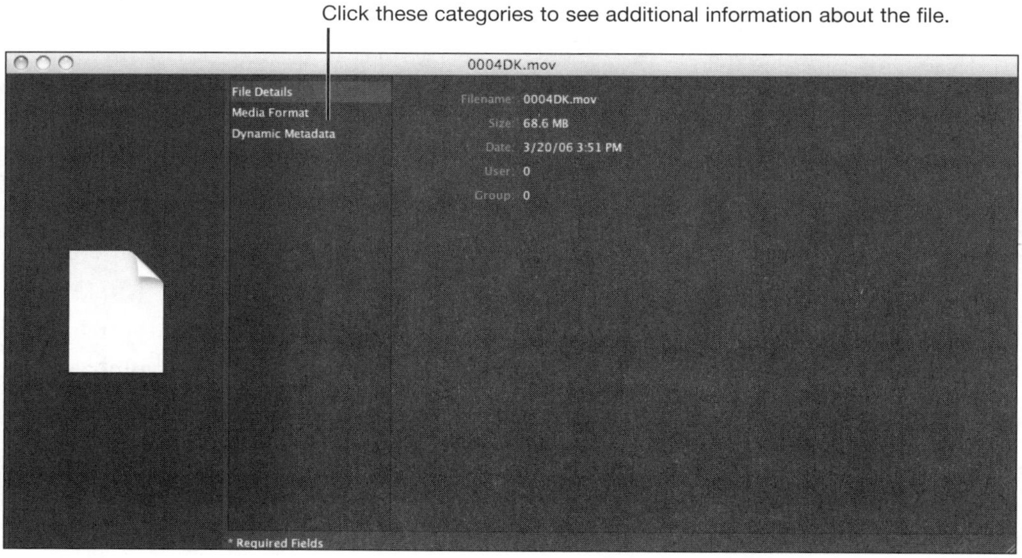

Viewing Files

To view a file, select View from the file's shortcut menu.

Because the file you may wish to view is not an asset, and therefore has no other representations such as a poster frame or a clip proxy, Final Cut Server will attempt to show you the file using the following process:

- It will cache the file by downloading it to the cache location on your computer. Remember, if it's a large file, this may take some time.
- After the file has been cached, Final Cut Server will try to open it using an application on your computer that has been assigned to open files like it. For example, on a Mac, text files will be opened with TextEdit, QuickTime files will open with QuickTime Player, and TIFF images will open with Preview.

Cataloging Device Files

If you wish to create an asset for a file, you're halfway home: The file is already on a recognized device, so there's no need to upload it. The only thing left to do is to catalog it.

There are two methods for cataloging a device file:

- Select Catalog from the file's shortcut menu.
- Select the file in the search results list and click the Catalog button in the device window's Toolbar.

Filename
00243K.mo

The Catalog window will appear.

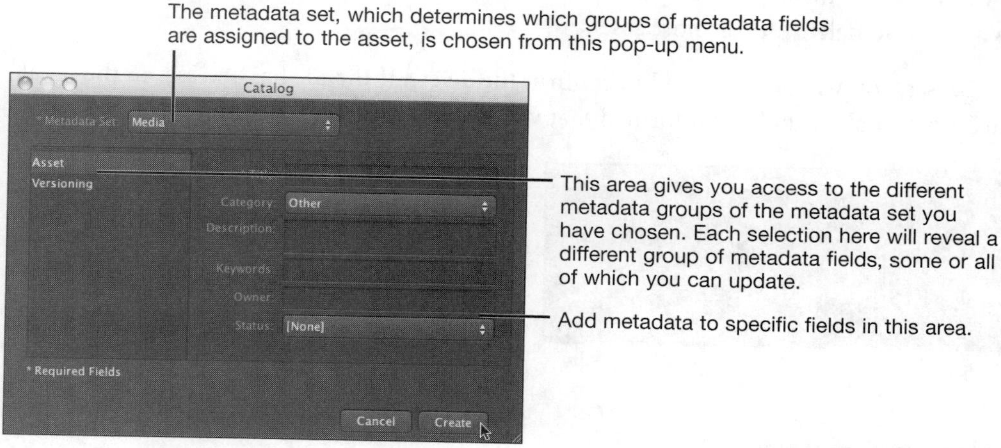
The metadata set, which determines which groups of metadata fields are assigned to the asset, is chosen from this pop-up menu.
Catalog
Media
Asset
Versioning
Category:
Other
Owner:
Status:
[None]
* Required Fields
Cancel
Create
This area gives you access to the different metadata groups of the metadata set you have chosen. Each selection here will reveal a different group of metadata fields, some or all of which you can update.
Add metadata to specific fields in this area.

Clicking the Create button allows you to search for the asset within the main client application window.

TIP Want to catalog more than one file at a time? No problem. Select them all and click the Catalog button to perform a Multiple Catalog. A special Multiple Catalog asset window will open in lieu of the one above. Its functionality is identical.

Exporting and Duplicating Assets from Devices

These functions operate exactly as other assets, as described in Lesson 2, "Working with Assets."

You start the process by Control-clicking the file in the search results list and selecting the appropriate function from the file's shortcut menu.

Deleting Files

Provided that your administrator has given you the access to perform the daunting task of deleting files from devices, you can do so in the Search Devices window.

Remember, you can't undo a delete function, so make sure you want to delete the file.

To delete a file, Control-click it in the search results list and choose Delete from its shortcut menu. To delete more than one file, select them together before opening the shortcut menu.

You can also use the keyboard shortcut Command-Delete (Mac) or Control-Delete (Windows) to delete one or more assets.

In all cases, you will be prompted to confirm the delete. If the file has an asset in the Final Cut Server catalog, you'll be informed that you'll be deleting the asset as well.

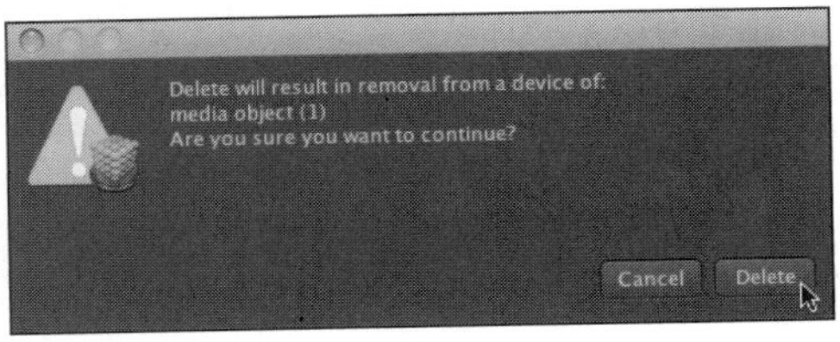
Delete will result in removal from a device of:
media object (1)
Are you sure you want to continue?
Cancel
Delete

Lesson Review

1. In the Search Devices window, what are the three things a user can search for?
2. Where can the Search Devices window be opened?
3. In the Search Devices window, users are able to search the contents of a device, but can you search for items that have not been added to the Final Cut Server catalog?

4. If an item has not been added to the Final Cut Server catalog, what are the two ways to catalog it in the Search Devices window?
5. True or false: You can catalog more than one file at a time.

Answers

1. File, still sequence, and composite clip.
2. This window can be opened from the Server pop-up menu in the main Final Cut Server application window.
3. You can search for all items of a device in the Search Devices window, regardless of whether the items have been added to Final Cut Server.
4. You can choose Catalog from the file's shortcut menu, or you can select the file in the search results list and click the Catalog button in the device window's Toolbar.
5. True. To add more than one item to the Final Cut Server catalog at a time, select all the items and click the Catalog button to perform a Multiple Catalog operation .

6

Goals

- View the Downloads & Uploads window
- Search for jobs
- View job details
- Retry and cancel jobs

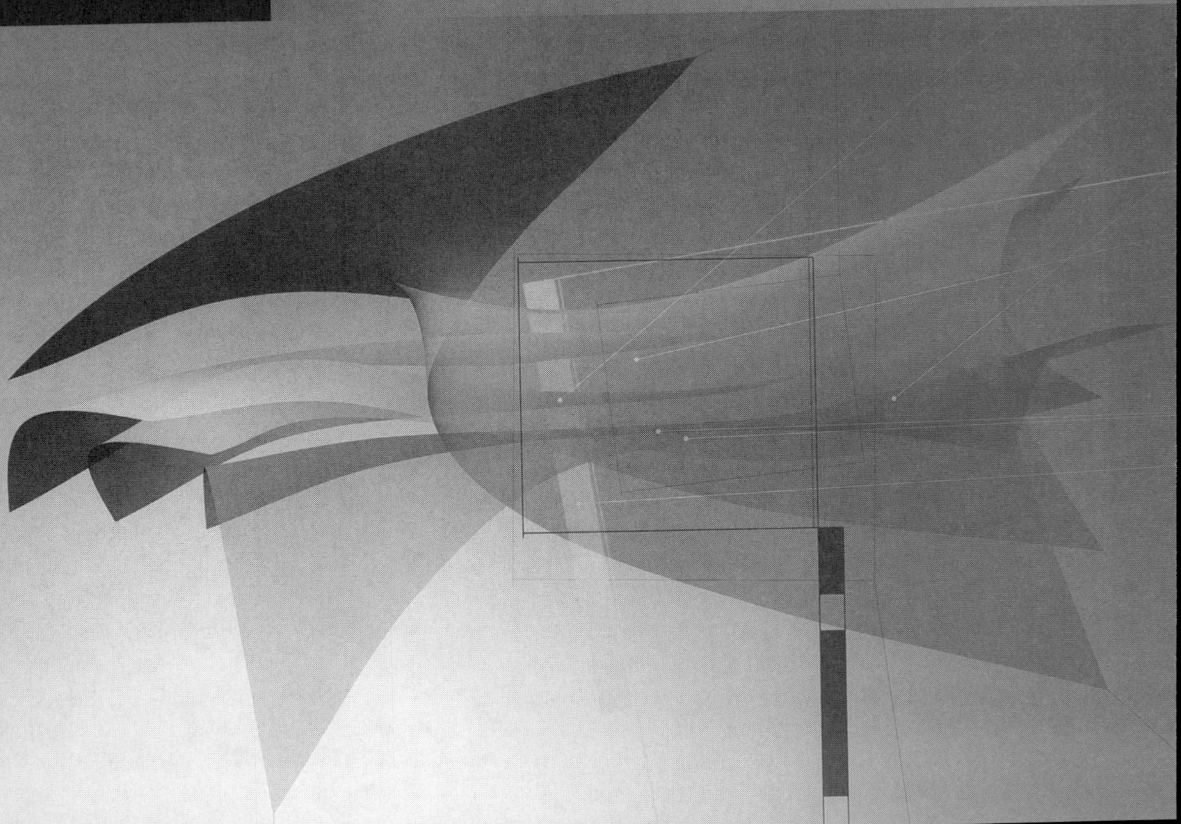

Lesson **6**

Working with Jobs

Every process that Final Cut Server does is a job, whether it's an upload, duplication, export, or transcode. The Downloads & Uploads window and the Jobs window both monitor these actions and provide information about them, such as successes or failures.

Viewing Downloads and Uploads

The Downloads & Uploads window shows only tasks that *you* have initiated. There are two methods for opening the Downloads & Uploads window:

- Click the Downloads & Uploads button in the bottom right of the main client application window.

- Choose Downloads & Uploads from the Window pop-up menu at the top left of the main client application window.

New Workspace
Downloads & Uploads

The Downloads & Uploads window appears.

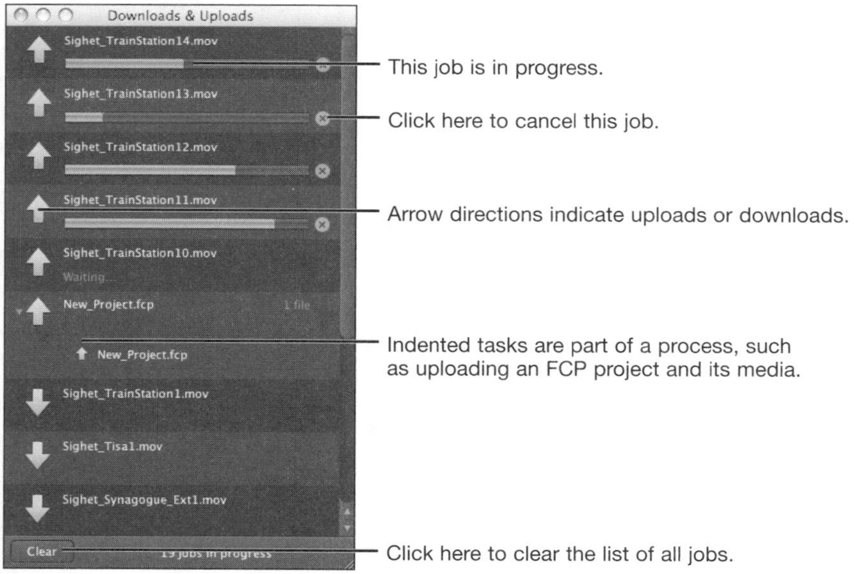
Downloads & Uploads
Sighet_TrainStation14.mov
Sighet_TrainStation13.mov
Sighet_TrainStation12.mov
Sighet_TrainStation11.mov
Sighet_TrainStation10.mov
New_Project.fcp
New_Project.fcp
Sighet_TrainStation1.mov
Sighet_Tisa1.mov
Sighet_Synagogue_Ext1.mov
Clear
This job is in progress.
Click here to cancel this job.
Arrow directions indicate uploads or downloads.
Indented tasks are part of a process, such as uploading an FCP project and its media.
Click here to clear the list of all jobs.

Viewing All Jobs

The Jobs window contains a list of all jobs that Final Cut Server is performing or has performed.

To open the Jobs window, choose Search All Jobs from the Server pop-up menu in the upper left of the client application window.

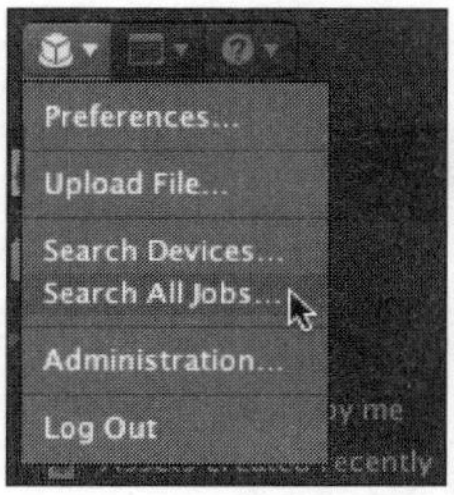
Preferences...
Upload File...
Search Devices...
Search All Jobs...
Administration...
Log Out

The Search All Jobs window appears.

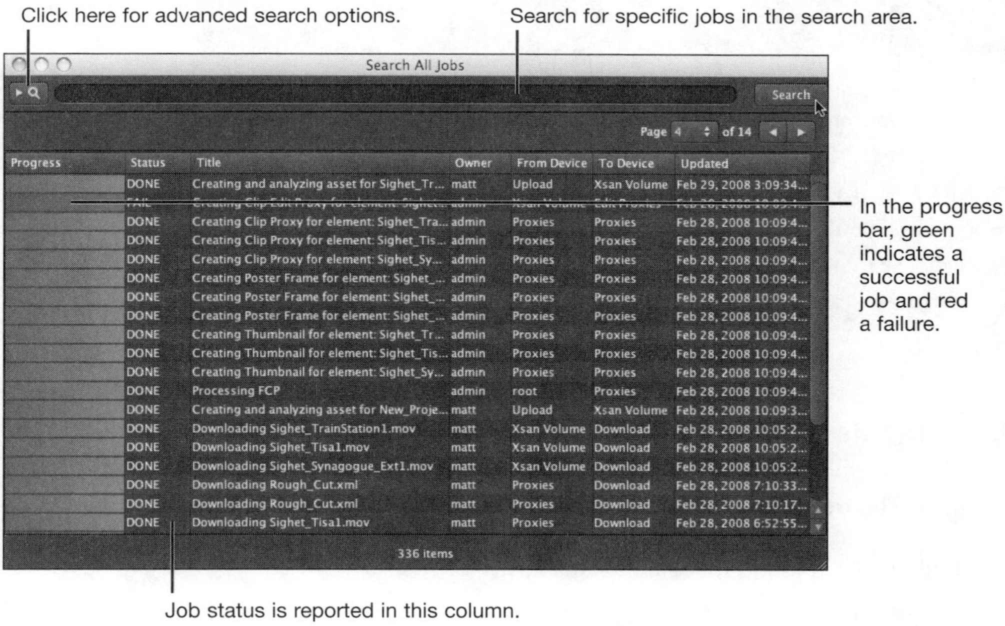
Click here for advanced search options.
Search for specific jobs in the search area.
Search All Jobs
Search
Page 4 of 14
Progress
Status
Title
Owner
From Device
To Device
Updated
336 items
In the progress bar, green indicates a successful job and red a failure.
Job status is reported in this column.

Searching for Jobs

You can search for jobs with specific criteria just as you would search for assets. Job searches have unique advanced search options.

The job log is usually archived periodically by your administrator. If you're searching for a job that occurred before this periodic cleanup time, then searching the job archive might yield the job you're looking for.

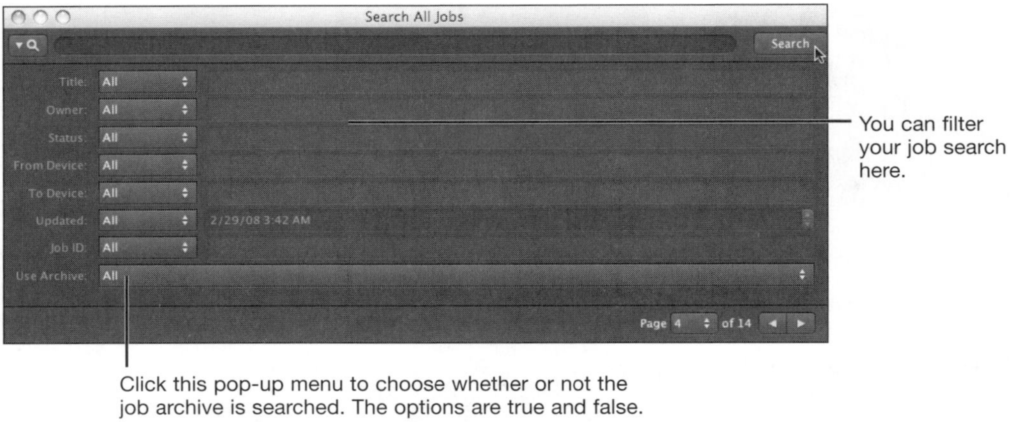

NOTE ▸ If your administrator has customized the layout of the advanced search options, you may see more or fewer fields than are pictured here. If you require additional search fields in your facility, talk to your administrator about customizing the advanced search options.

Viewing Job Details and Error Messages

The job info window has lots of information about a job, as well as specific reasons why a job failed. There are two methods for opening a job info window:

- Double-click a job in the search results list.
- Choose Get Info from the job's shortcut menu.

The job info window appears.

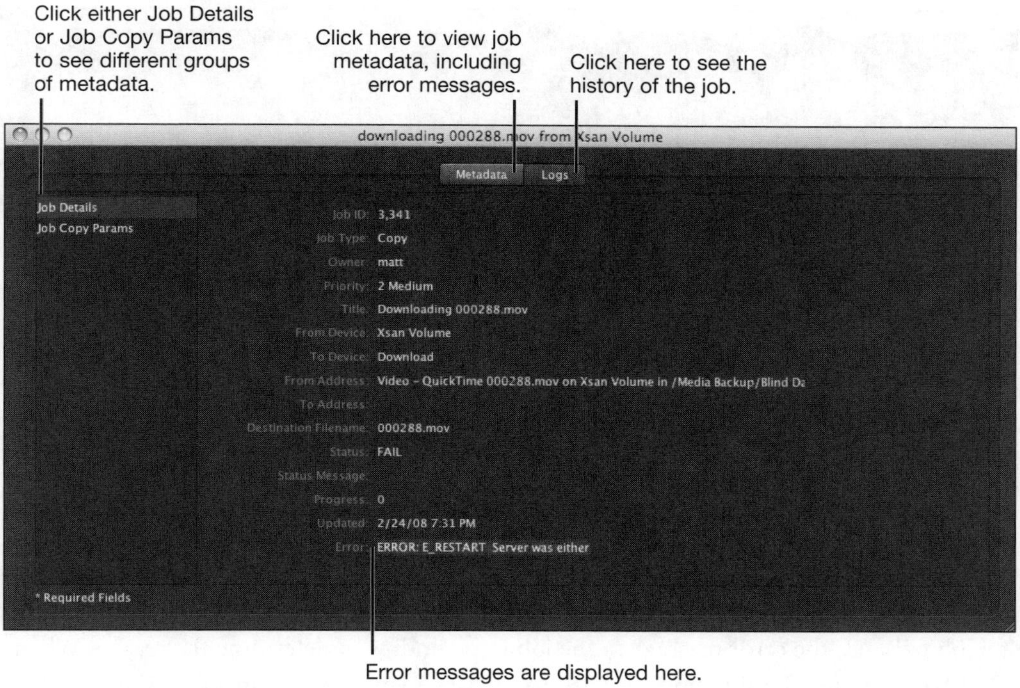

When viewing metadata in this window, you may see a partial statement, like that shown in the error field in this screen shot. To see the entire message, drag your mouse within the field.

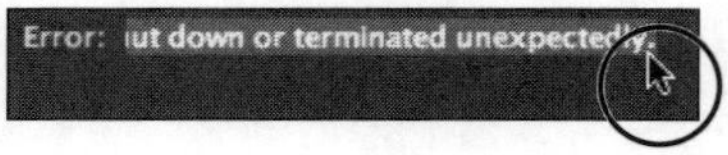

The Logs pane displays a full history of the job.

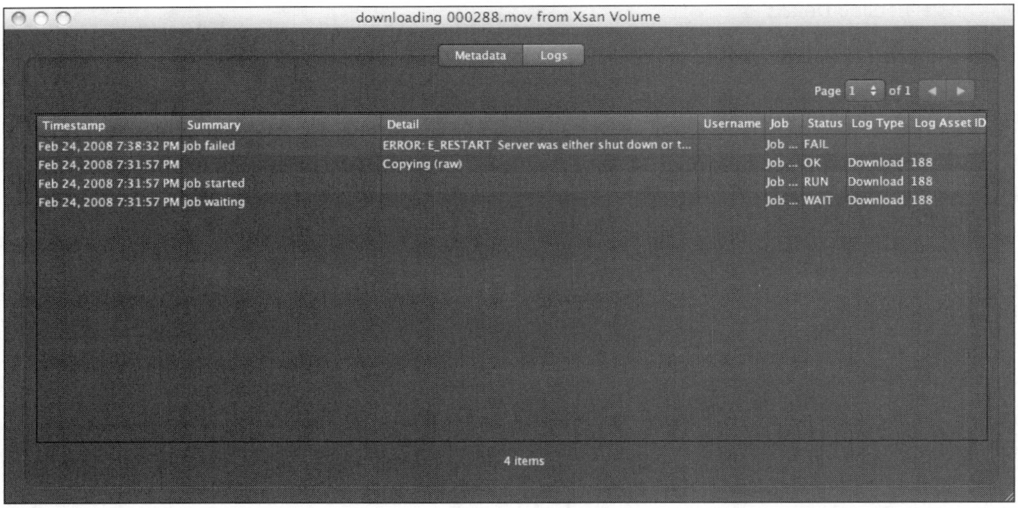

Retrying and Canceling Jobs

If a job fails, use the error message in the job info window to determine the reason. When you have corrected the situation (or tried to, anyway), you can tell Final Cut Server to retry the job.

Control-click a job to bring up its shortcut menu and choose Retry.

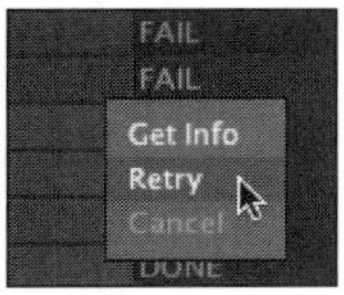

NOTE ▶ You can only retry jobs that you initiated, such as uploads or exports. Ask your administrator to retry or troubleshoot other failed jobs.

Similarly, you can choose Cancel to stop a currently running job. The Cancel option will be dimmed if a job has failed or completed, or if you don't have access to cancel it.

Lesson Review

1. As a user (not an administrator), if a job failed, where would you find out details about its error?
2. True or false: The Search All Jobs window can be accessed from the Toolbar of the main client window.
3. In terms of clip proxy files, how does the Downloads & Uploads window differ from the Search All Jobs window?
4. How do you show a list of only your failed jobs?
5. Name four of the progress and status definitions.

Answers

1. You can get job details and error messages either by choosing Get Info from the job's shortcut menu or by double-clicking the failed job. The error messages are in both the Metadata and Logs tabs of the Job Details section.
2. False. The Search All Jobs window can be accessed from the Final Cut Server pop-up menu in the main Final Cut Server window.
3. The Downloads & Uploads window does not show details about files being transcoded to create clip proxies.
4. Type *fail* in the search tab of the Search All Jobs window and click Search.
5. Done, run, wait, and fail are the possible progress and status definitions.

Basic Final Cut Server Administration

7

Goals

Set up groups of users to log in

Define and configure devices

Create file system watchers

Create asset metadata subscriptions

Back up the Final Cut Server catalog

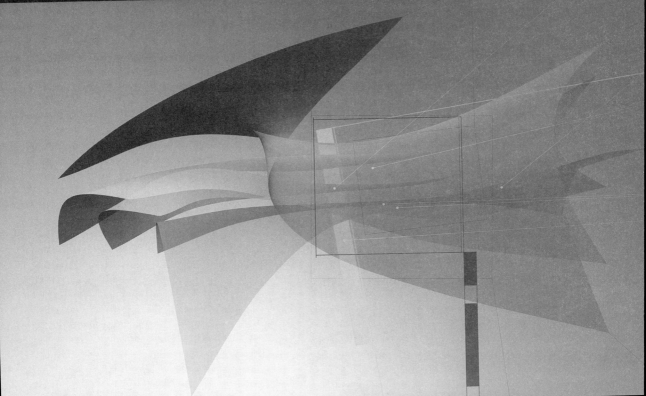

Lesson **7**

Basic Administration

If you've flipped to this section of the book, willingly or unwillingly, you are probably the person assigned the task of administering your organization's Final Cut Server.

Although Final Cut Server is useful immediately after installation, it really starts to shine with the customization of features and the creation of automations.

Rest assured, administering your Final Cut Server can be quite simple when accomplished through the Final Cut Server System Preferences pane, which is covered in this lesson.

For an exploration of Final Cut Server's advanced administration offerings, such as the customization of metadata or for more precise control over automations, skip on to Lesson 8, "Advanced Administration."

> **NOTE ▶** For instructions on how to install Final Cut Server, please refer to the Appendix, Best Practices for Installation. Also, a tutorial is available on Apple's website, and Lesson 2 of the Setup and Administration Guide, which comes with Final Cut Server, gives straightforward instructions as well.

Accessing Administration Tools

The Final Cut Server System Preferences pane gives you access to the most often customized features of the program. When you delve into certain tasks, assistants are launched to help guide you through the process even further.

NOTE ► The Final Cut Server System Preferences pane is located only on your Final Cut Server. In order to access the preference pane, you must be viewing a display hooked up to your Final Cut Server, or be controlling it through Leopard Screen Sharing or Apple Remote Desktop.

The preference pane is located in System Preferences, in the Other category.

Clicking it reveals the preference pane open to the General tab.

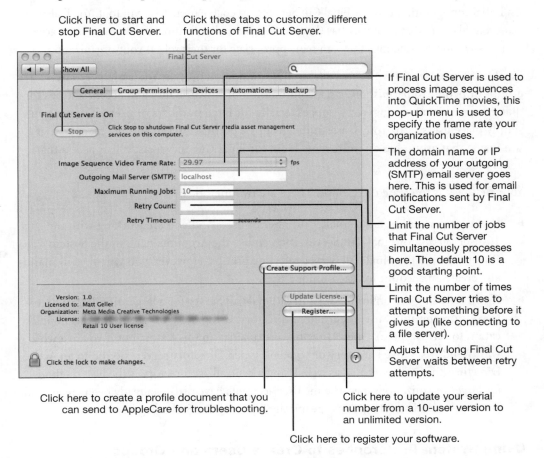

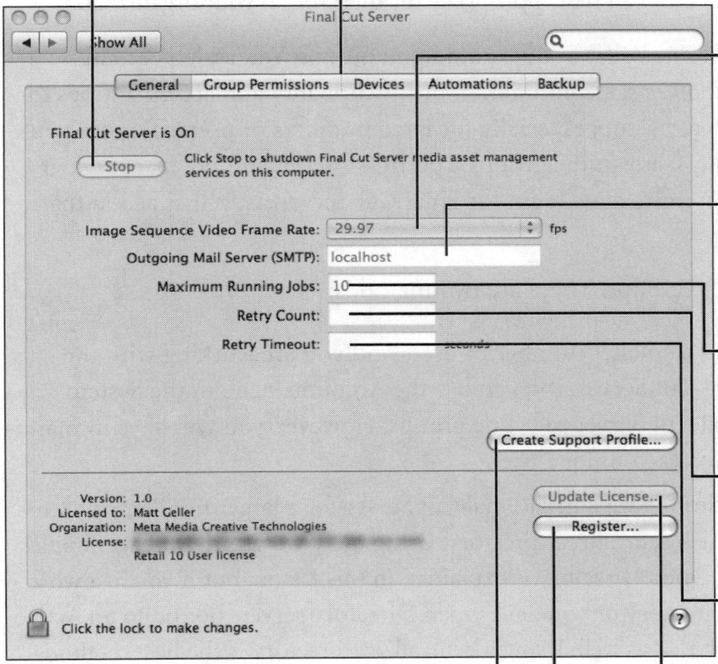
Click here to start and stop Final Cut Server.
Click these tabs to customize different functions of Final Cut Server.
Final Cut Server
Show All
General
Group Permissions
Devices
Automations
Backup
Final Cut Server is On
Stop
Click Stop to shutdown Final Cut Server media asset management services on this computer.
Image Sequence Video Frame Rate: 29.97 fps
Outgoing Mail Server (SMTP): localhost
Maximum Running Jobs: 10
Retry Count:
Retry Timeout:
Create Support Profile...
Version: 1.0
Licensed to: Matt Geller
Organization: Meta Media Creative Technologies
License:
Retail 10 User license
Update License...
Register...
Click the lock to make changes.
If Final Cut Server is used to process image sequences into QuickTime movies, this pop-up menu is used to specify the frame rate your organization uses.
The domain name or IP address of your outgoing (SMTP) email server goes here. This is used for email notifications sent by Final Cut Server.
Limit the number of jobs that Final Cut Server simultaneously processes here. The default 10 is a good starting point.
Limit the number of times Final Cut Server tries to attempt something before it gives up (like connecting to a file server).
Adjust how long Final Cut Server waits between retry attempts.
Click here to create a profile document that you can send to AppleCare for troubleshooting.
Click here to update your serial number from a 10-user version to an unlimited version.
Click here to register your software.

As a general rule, in order to make changes to the settings in any preference pane, you will have to click the lock at the lower left, then authenticate with your administrator username and password.

Defining Groups and Permission Sets

An initial configuration task for Final Cut Server is to enable users to log in to it. You'll also want to make sure that each user has the functionality (what features they can use) and access (what assets and productions they can see) appropriate for their role at your organization.

Final Cut Server works with one collaborative paradigm: groups. You assign certain groups of people to *permission sets*, a combination of functionality and access settings for that group. This makes for easy setup, especially for large numbers of users, because you simply place users into groups, assign the groups a permission set, and *presto*, every person inside that group has an immediate login for Final Cut Server, as well as access that befits their role.

Your first task is to create the groups. There are two methods:

- If your organization has a smaller number of users and you are working with the 10-user version of Final Cut Server, you can use the Accounts pane in the System Preferences on the Final Cut Server to create groups. However, you will have to manually create an account for each unique user.
- If you are using the unlimited version of Final Cut Server for a larger number of users in your organization, use a centralized directory. Using the Workgroup Manager application to access Apple's Open Directory is explained in this lesson, but if you're a whiz with Windows Active Directory or you have Open Directory servers bound to an Active Directory, that will work just as well. Using a centralized directory is the best method for larger organizations, because the user accounts have usually already been created. All you have to do is create groups specifically for Final Cut Server login.

Using System Preferences to Create Users and Groups

Manually entering users and groups in System Preferences is best for small shops that are using the 10-user license version of Final Cut Server and don't have a centralized directory.

1 On the Final Cut Server, go to System Preferences. If you're already there, you might need to click the Show All button in the upper right to see all of the preference panes.

2 Click the Accounts icon.

3 In the Accounts pane, click the plus (+) button, at bottom left, to create a new user.

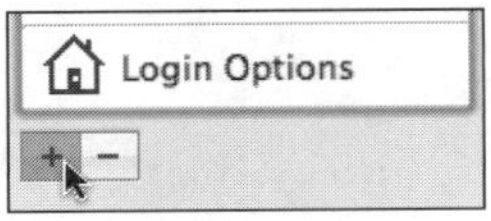

4 Create a username and password for the user, and click Create Account.

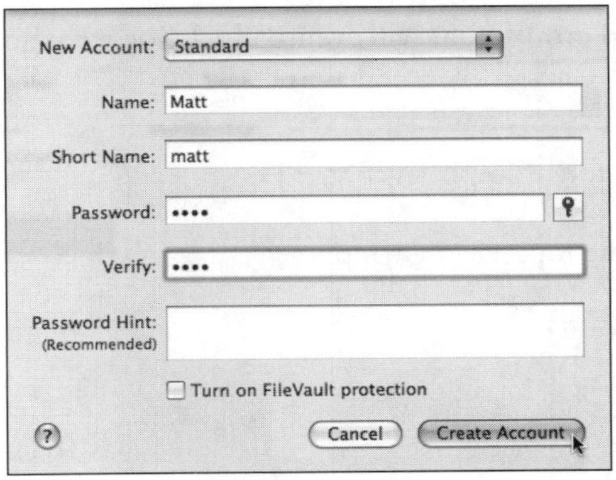

5 If this is the first user you are creating on this system (besides your original admin user), you will be asked if you want to turn off automatic login.

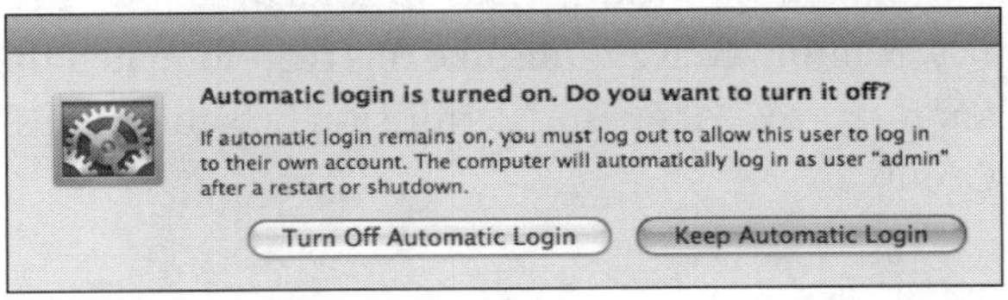

The users you are creating for the Final Cut Server don't log in to this machine in the traditional sense, so you might be tempted to keep the automatic login on. However, as the Final Cut Server becomes an important part of your organization's workflow, it will be important to secure it.

Turn off automatic login to ensure that only people with administrator rights can log in to the Final Cut Server.

You will now see the user listed in the column at left.

Repeat the steps to create all the users you will need.

Defining Groups

The next step is to combine these users into groups that will eventually define what access and functionality rights they have.

1 To make a group, click the plus (+) button at lower left.

2 From the New Account pop-menu at the top, select Group.

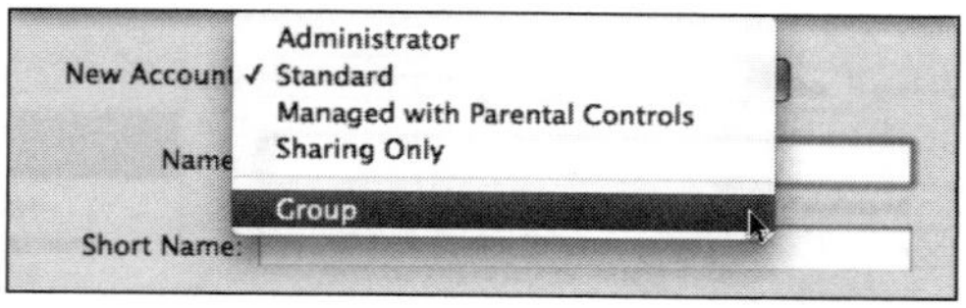

A new sheet drops down, asking for a name for the group.

3 In the Name field, name the group something synonymous with its function, such as Producers, Editors, or Graphics. Naming conventions like this help to match the group to its appropriate permission set (see "Assigning Permission Sets to Groups" in this lesson). Then click Create Group.

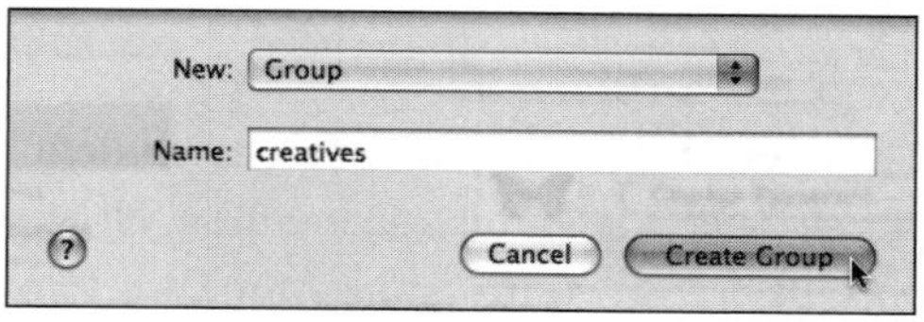

4 In the new pane that appears at the right of the Accounts pane, assign users to a group by selecting the checkboxes to the left of their names.

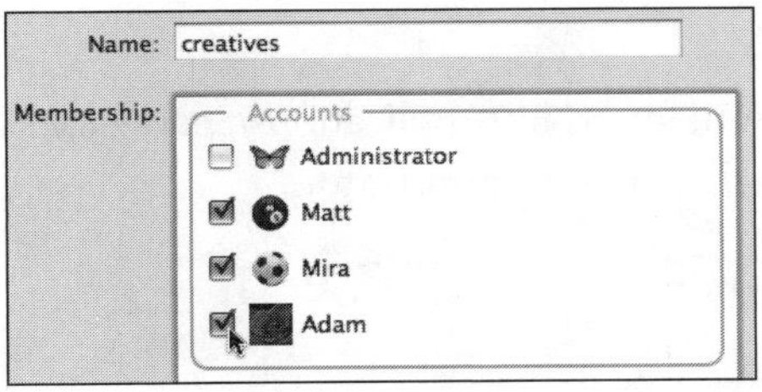

NOTE ► As long as you originally installed the software while logged in with your admin user account, there's no need to place your admin user account in one of these groups. Your admin user account always has the ability to log in to Final Cut Server and has full privileges.

Create as many groups as fit the roles at your organization. To use them in Final Cut Server, skip over the next section and pick up with "Adding Groups to Final Cut Server."

Using Workgroup Manager to Create Groups

If you already use Apple's Open Directory, you can follow this method. Before proceeding, make sure you've met these three criteria:

- The Open Directory has been created, it is running well, and users have already been entered into it.
- The Final Cut Server is "bound" to the Open Directory.
- You have full directory administrator access to the Workgroup Manager program, which is used to access the Open Directory system.

MORE INFO ► There are great resources for this topic, including *Apple Training Series: Mac OS X v10.5 Directory Services*, by Arek Dreyer (Peachpit Press, 2008), if you need help with any of the above.

All you need to do is create groups. You can even get away with using current groups if they contain some of the users you want to assign to a specific permission set. Remember that adding additional users to the group now or later will automatically enable those users to log in as well.

If your directory already has the groups you need, you don't really need to add new ones and can skip to the next section. The following steps assume that you're using Apple's Open Directory and need to create new groups.

1 Launch the Workgroup Manager application, found in /Applications/Server, and log in to your Open Directory Master as your directory administrator.

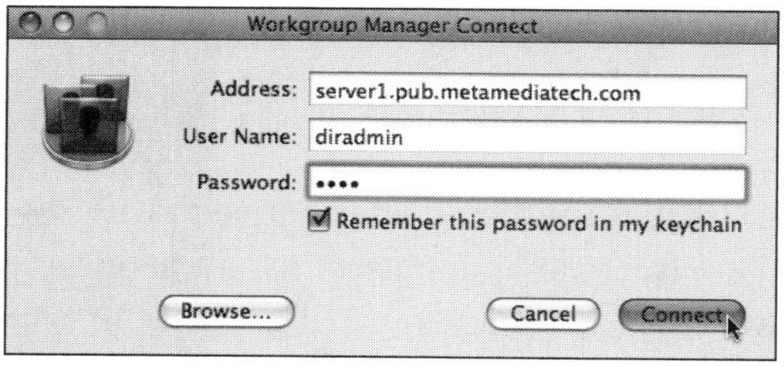

The main Workgroup Manager window appears.

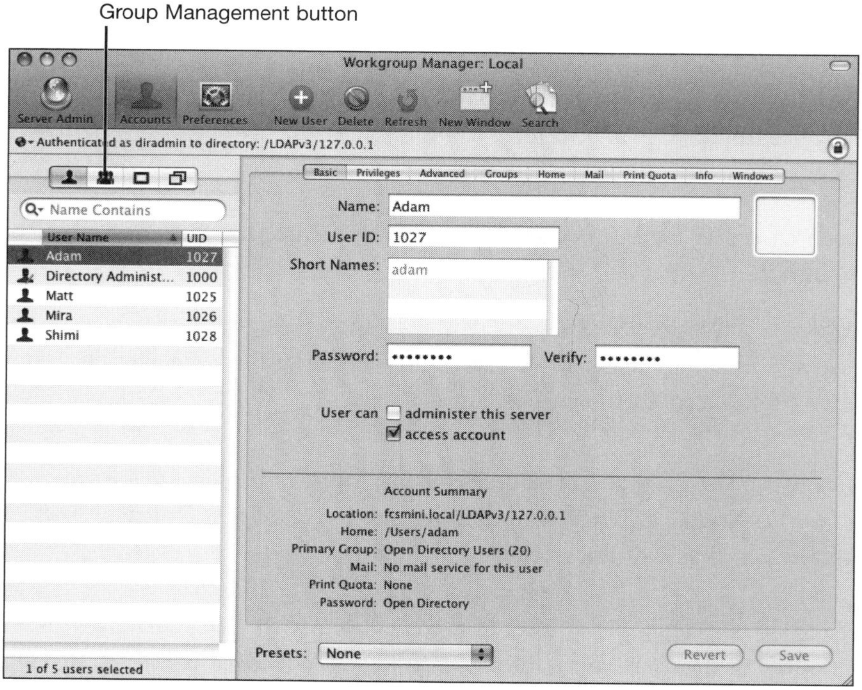

2 In the upper left of the interface, click the Group Management button.

The Group Management pane appears, and the toolbar in the upper right changes to show tools specific to group management, including the New Group button.

3 Click the New Group button.

4 Name the group something indicative of its function, so that mapping the permission set in Final Cut Server will be easy to remember.

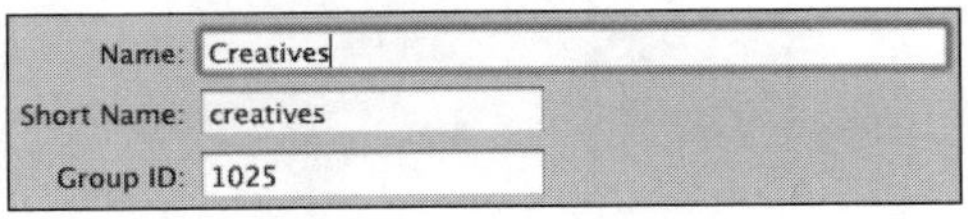

5 Click the Members tab at right to add users to this new group.

The Members pane opens.

6 In the Members pane, click the plus (+) button to reveal a list of the current users.

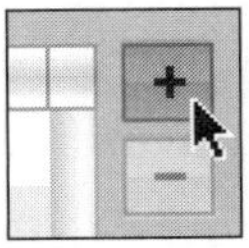

7 Drag and drop the appropriate users into the Members area.

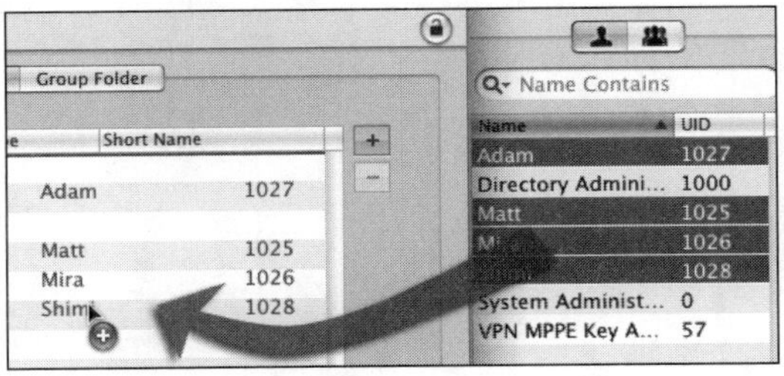

Once you save the changes and quit Workgroup Manager, you're ready to add the group(s) to Final Cut Server.

Adding Groups to Final Cut Server

After creating your groups (using either System Preferences or Workgroup Manager), return to the Final Cut Server System Preferences pane, and click the Group Permissions tab.

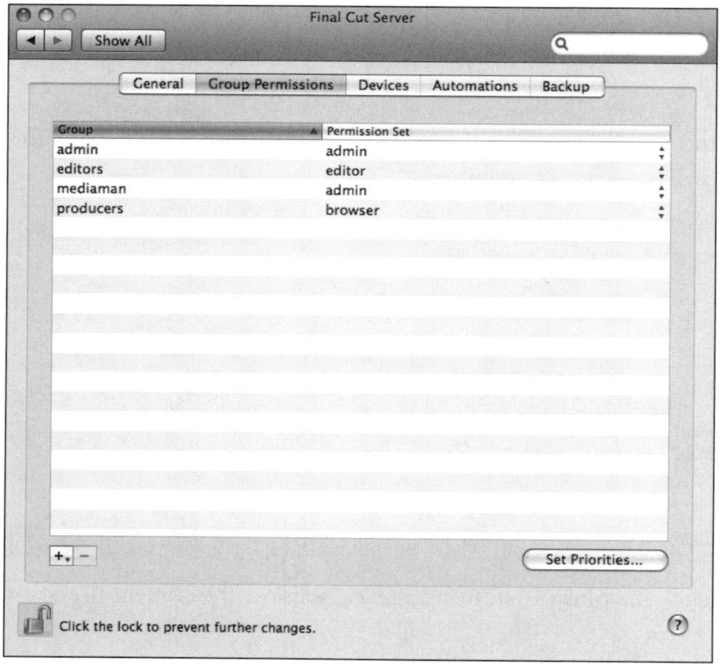

Click the plus (+) button at lower left to add the groups you have created into the list.

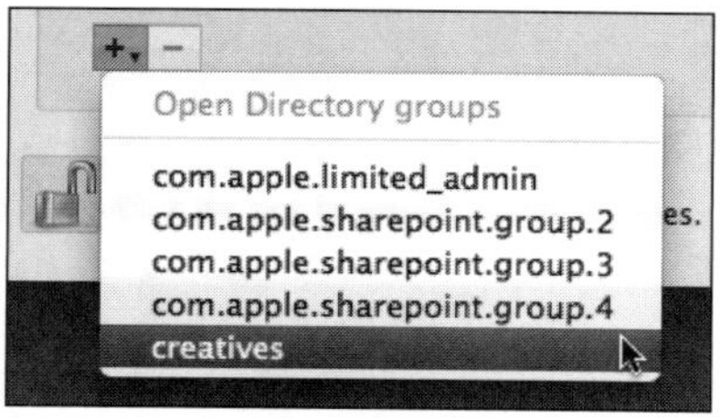

Assigning Permission Sets to Groups

Once the groups have been selected, they will show up in the main window with a blank permission set assignment. Simply click in the blank area and select the appropriate permission set.

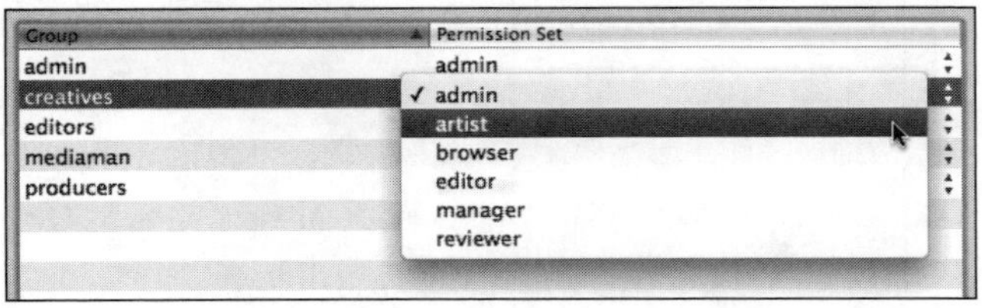

Your list of permission sets depends on the customer profile you chose during installation. Detailed descriptions of the default permission sets can be found in the Final Cut Server Setup and Administration Guide. Most permission set names are intuitive, but it's a good idea to check out their specific definitions to ensure that you are picking the correct one.

You may want to create new permission sets, or tweak the functionality and access rights of existing ones. All of this is done (like many of the more granular administrative tasks within Final Cut Server) inside the advanced administration section of the client application. More on this is revealed in Lesson 8, "Advanced Administration."

Setting Group Priorities

If a user is a member of more than one group, the group with highest priority will determine the user's permission set. By default, Final Cut Server orders groups so that those with broader permission sets have higher priority, thus ensuring the greatest possible access to a user. Final Cut Server displays the permission set with highest access priority at the top of the list and the lowest at the bottom.

If your situation is unique, you can reorder them:

1 In the Group Permissions pane, click the Set Priorities button, at bottom left.

2 In the dialog that opens, order the permission sets to your needs using drag and drop. Click OK when done.

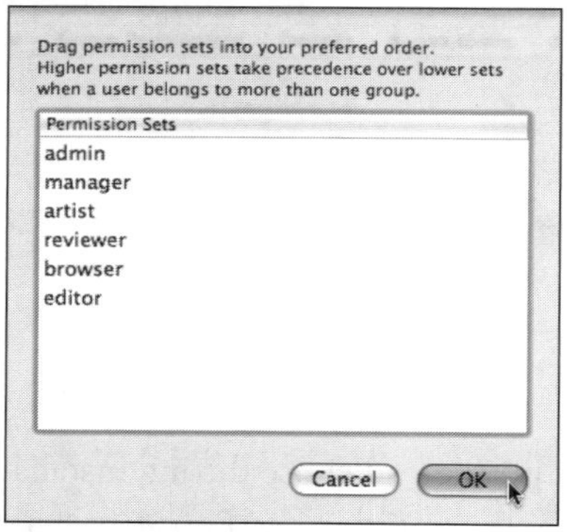
Drag permission sets into your preferred order.
Higher permission sets take precedence over lower sets
when a user belongs to more than one group.
Permission Sets
admin
manager
artist
reviewer
browser
editor
Cancel
OK

Checking Logins and Permissions

An important final step here is to log in as several users and check to see if Final Cut Server accepts the login. Does the user experience within the client application match your intentions for the user's group? You should prepare a few simple test processes to verify each user's experience on different systems. Instructions on how to download the client application are given in Lesson 1, "Overview and Interface Basics."

Defining Devices

The next major task is to make sure that Final Cut Server "sees" all of the storage systems that your organization uses. These storage systems are called *devices* in Final Cut Server, and can mean anything from storage connected directly to the Final Cut Server, such as a FireWire drive, to the Xsan volumes in your shop, to AFP or SMB file servers in another department, and even FTP servers in China.

Determining Device Requirements

Because a lot of these devices will already have folders and files on them, a decision has to be made whether the entire storage system, or a specific folder inside the system, should be a device.

For example, if an AFP file server has only one folder that stores files pertinent to your workflow, then *that folder* should be a device, and not the entire AFP file server.

Another question to ponder is whether you wish Final Cut Server to regularly *scan* the device for new or modified media. Obviously, if files already exist on a device and they need to be cataloged, then scanning is a great way to ensure that all of those files become assets.

Finally, you can also create *archive devices,* which are devices dedicated to archiving assets that are no longer being used. Archive devices can be any kind of storage system mentioned above. They are only used to store the primary representations (hi-res files) through an archive process that is triggered by an automation you configure or by an end user who has the right to archive assets. Ideal archive devices are *hierarchical storage management systems,* or HSMs. These storage systems behave like black holes, in that they can continuously accept files and then, according to predefined rules, transfer them to more stable storage media, like tape. These kinds of systems must be preconfigured and managed separately from Final Cut Server.

HSMs present archived items as if they were still in their original archive location. This is an important feature. If Final Cut Server ever needs to restore a file to its original location, it relies on the primary representation being in the same place on an archive device. If it isn't, then you the administrator have to get it there yourself. For example, if a FireWire drive was configured to be an archive device, and you have since disconnected it because it was full, you'll have to go get it and plug it back into the Final Cut Server if you ever have to restore files from it.

There are many ways to add and manage devices. We will only be looking at a few options in this section.

Adding a New Device

Once you have determined the needs of your organization, you can begin the process of adding devices to your Final Cut Server.

1 Click the Devices tab in the Final Cut Server System Preferences pane.

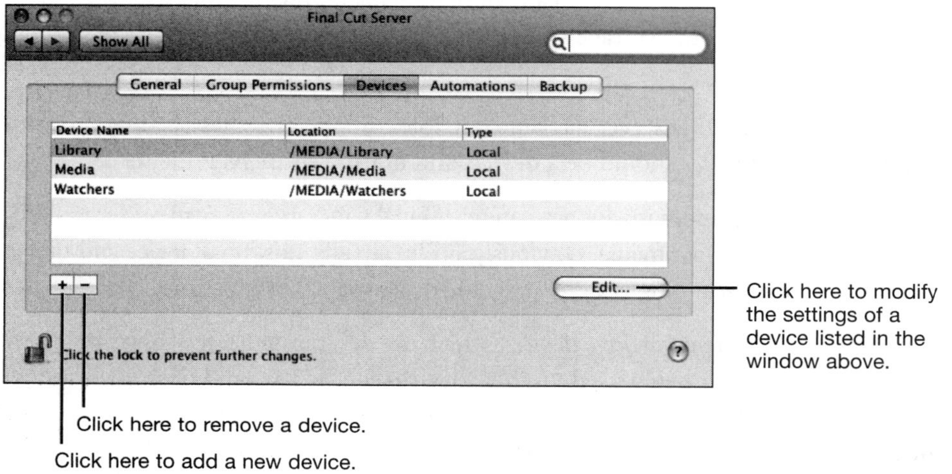
Final Cut Server
Show All
General
Group Permissions
Devices
Automations
Backup
Device Name
Location
Type
Library
/MEDIA/Library
Local
Media
/MEDIA/Media
Local
Watchers
/MEDIA/Watchers
Local
Edit...
Click the lock to prevent further changes.
Click here to modify the settings of a device listed in the window above.
Click here to remove a device.
Click here to add a new device.

2 Click the plus (+) button to launch the Device Setup Assistant.

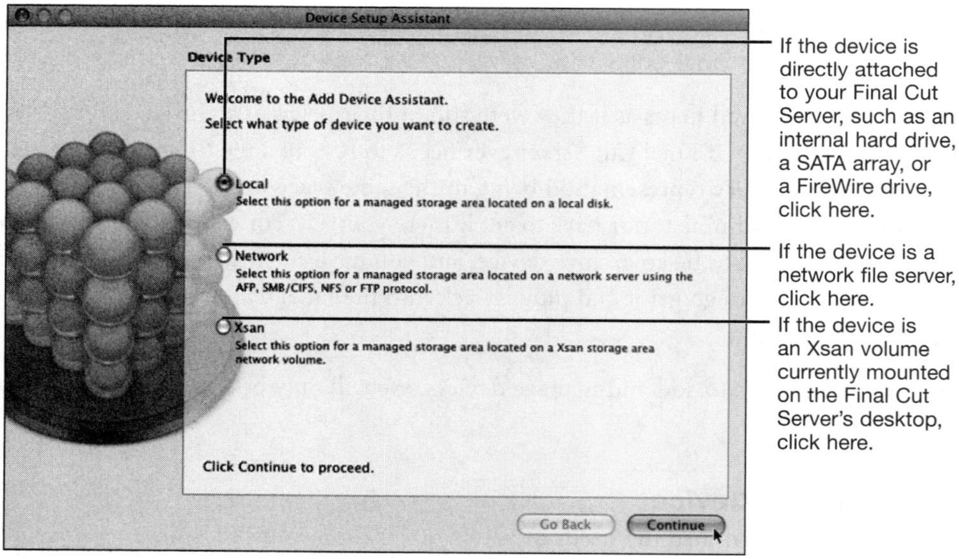
Device Setup Assistant
Device Type
Welcome to the Add Device Assistant.
Select what type of device you want to create.
Local
Select this option for a managed storage area located on a local disk.
Network
Select this option for a managed storage area located on a network server using the AFP, SMB/CIFS, NFS or FTP protocol.
Xsan
Select this option for a managed storage area located on a Xsan storage area network volume.
Click Continue to proceed.
Go Back
Continue
If the device is directly attached to your Final Cut Server, such as an internal hard drive, a SATA array, or a FireWire drive, click here.
If the device is a network file server, click here.
If the device is an Xsan volume currently mounted on the Final Cut Server's desktop, click here.

3 The next pane will differ depending on the selection you made. Local devices need a name and a specific file path. Remember, the root of the volume might not be the best place to define for the device.

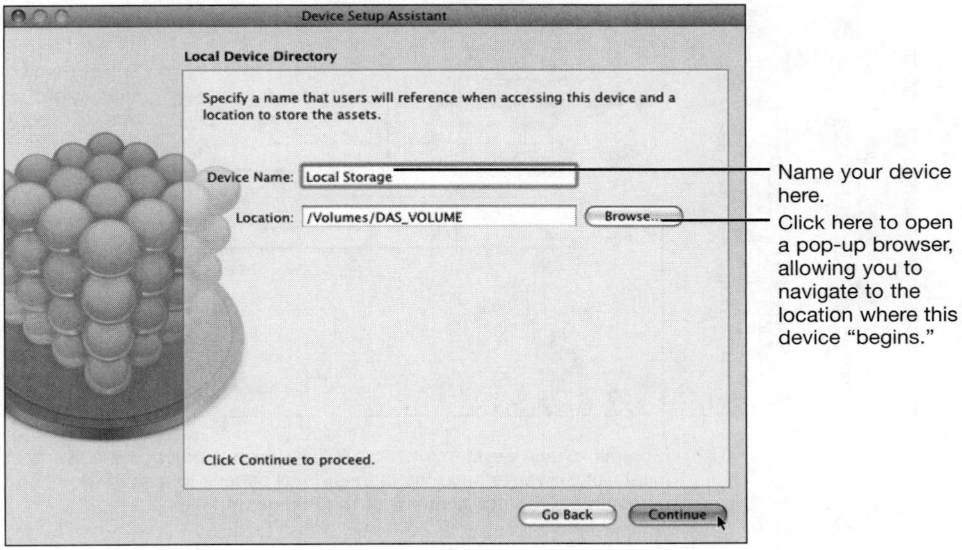
Device Setup Assistant
Local Device Directory
Specify a name that users will reference when accessing this device and a location to store the assets.
Device Name: Local Storage
Location: /Volumes/DAS_VOLUME
Browse...
Click Continue to proceed.
Go Back
Continue
Name your device here.
Click here to open a pop-up browser, allowing you to navigate to the location where this device "begins."

Xsan devices are set up with a nearly identical window. The configuration is the same, although the assistant will perform a check before proceeding to make sure it is indeed an Xsan volume.

Network devices have a different configuration window.

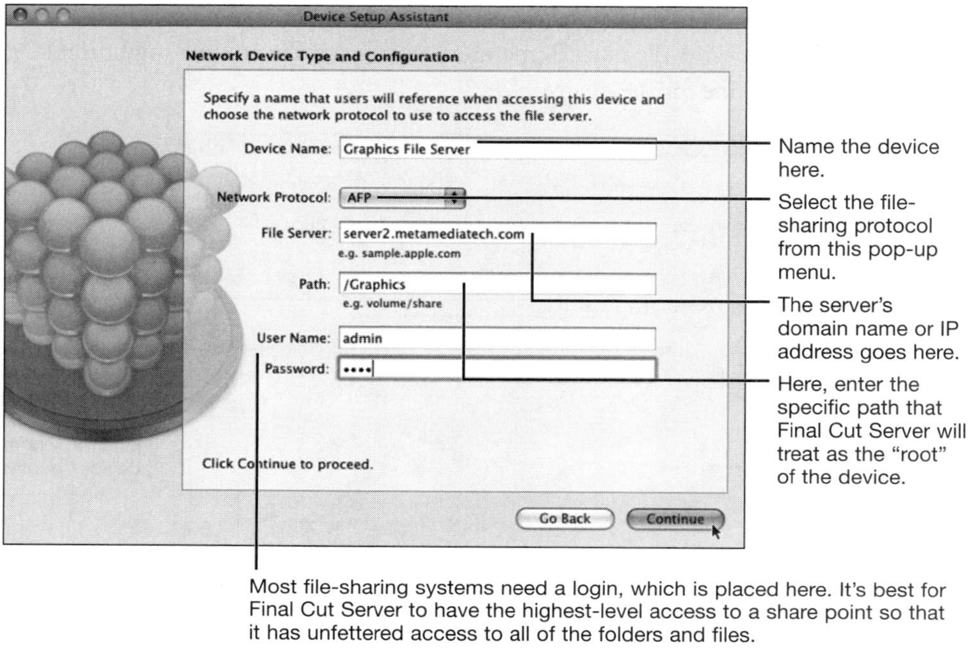
Device Setup Assistant
Network Device Type and Configuration
Specify a name that users will reference when accessing this device and choose the network protocol to use to access the file server.
Device Name: Graphics File Server
Network Protocol: AFP
File Server: server2.metamediatech.com
e.g. sample.apple.com
Path: /Graphics
e.g. volume/share
User Name: admin
Password: ••••
Click Continue to proceed.
Go Back
Continue
Name the device here.
Select the file-sharing protocol from this pop-up menu.
The server's domain name or IP address goes here.
Here, enter the specific path that Final Cut Server will treat as the "root" of the device.
Most file-sharing systems need a login, which is placed here. It's best for Final Cut Server to have the highest-level access to a share point so that it has unfettered access to all of the folders and files.

The Network Protocol pop-up menu allows you to choose from four different kinds of file servers.

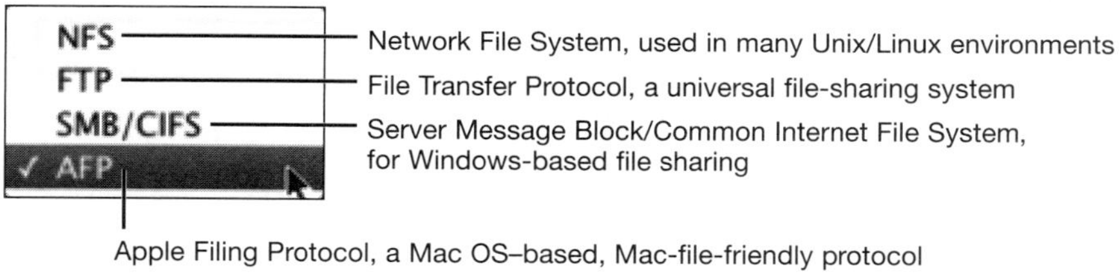
NFS
FTP
SMB/CIFS
✓ AFP
Network File System, used in many Unix/Linux environments
File Transfer Protocol, a universal file-sharing system
Server Message Block/Common Internet File System, for Windows-based file sharing
Apple Filing Protocol, a Mac OS–based, Mac-file-friendly protocol

4 The next window simply asks if the device should be an archive device.

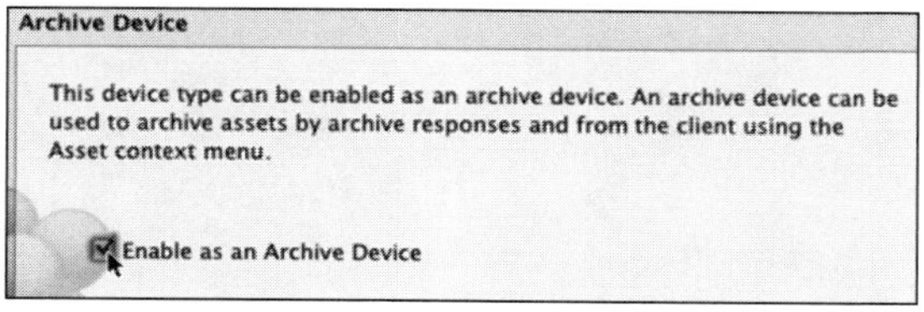
Archive Device
This device type can be enabled as an archive device. An archive device can be used to archive assets by archive responses and from the client using the Asset context menu.
Enable as an Archive Device

To make the device an archive, select Enable as an Archive Device and then click Continue. At that point, you're pretty much done.

If this device is meant for other endeavors, leave the box blank and click Continue to access the Scan Settings window.

5 In the Scan Settings window, you can turn on scanning, configure its frequency, and select a metadata set for uploads.

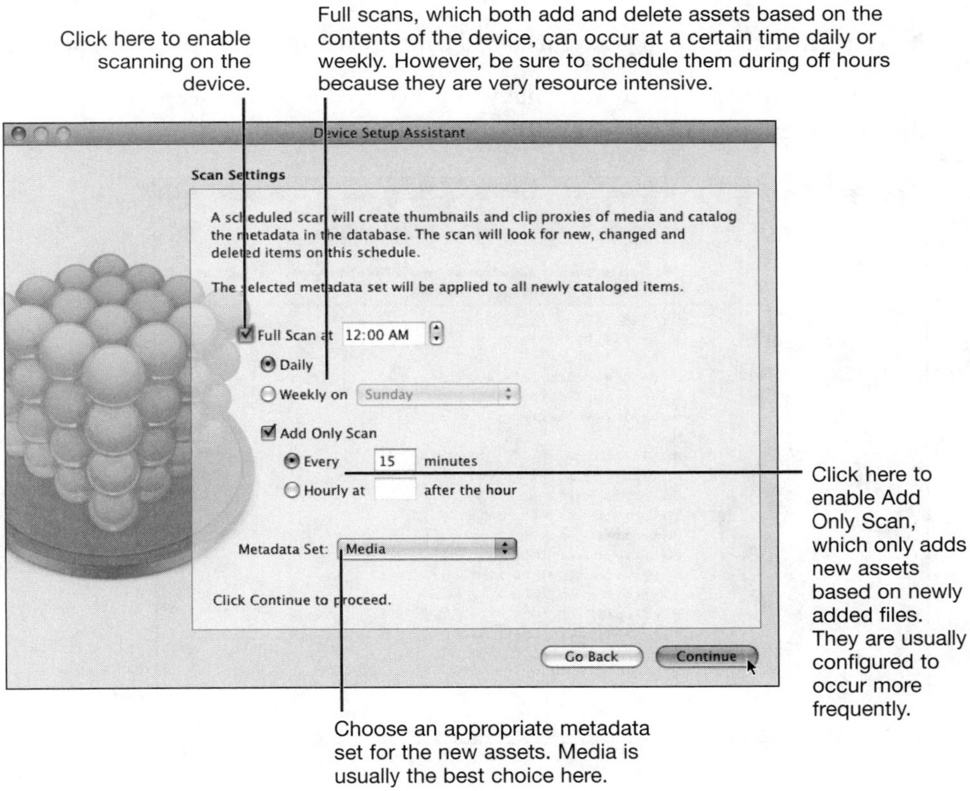
Click here to enable scanning on the device.
Full scans, which both add and delete assets based on the contents of the device, can occur at a certain time daily or weekly. However, be sure to schedule them during off hours because they are very resource intensive.
Device Setup Assistant
Scan Settings
A scheduled scan will create thumbnails and clip proxies of media and catalog the metadata in the database. The scan will look for new, changed and deleted items on this schedule.
The selected metadata set will be applied to all newly cataloged items.
Full Scan at 12:00 AM
Daily
Weekly on Sunday
Add Only Scan
Every 15 minutes
Hourly at after the hour
Metadata Set: Media
Click Continue to proceed.
Go Back
Continue
Click here to enable Add Only Scan, which only adds new assets based on newly added files. They are usually configured to occur more frequently.
Choose an appropriate metadata set for the new assets. Media is usually the best choice here.

After choosing your scan settings, click Continue.

6 In the next window, you are asked to pick transcode settings for the device. The selections here determine what formats should be chosen to convert files when uploading them to this device or duplicating them from another device.

For example, if your organization wants all video files on an FTP site to be encoded in H.264, you can choose this as your transcode setting. All jobs, whether submitted automatically or manually, can be converted to that format.

You can select as many or as few settings as you wish. Just remember these ideas:

- The fewer choices you give your staff, the more likely they'll choose the right setting.
- Unless you're absolutely sure that you won't need it, always keep No Conversion, the first selection in the list, checked as an option.

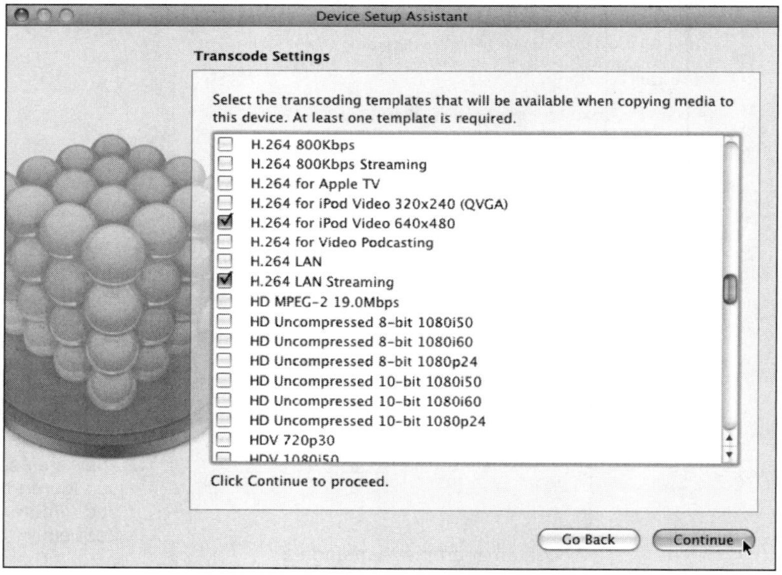

7 A final summary window displays the cumulative choices you made about this device.

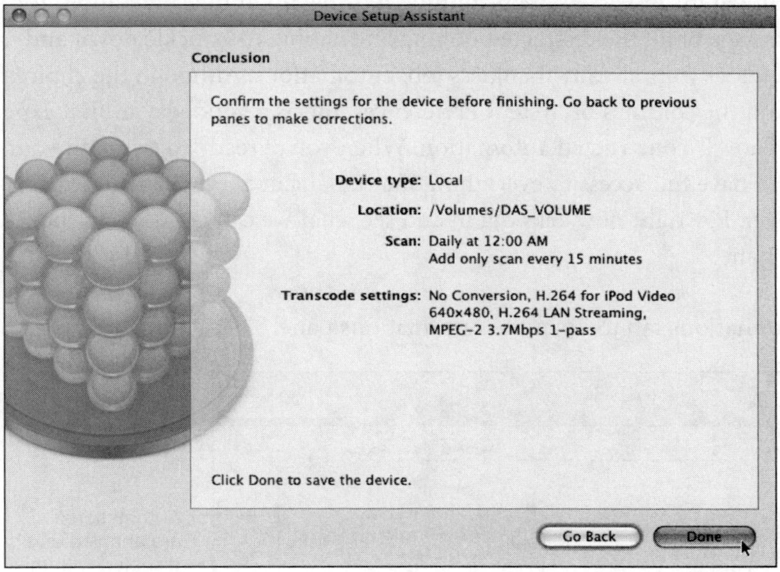

8 Click Done; your new device is now listed among the others.

Device Name	Location	Type
Library	/MEDIA/Library	Local
Media	/MEDIA/Media	Local
Watchers	/MEDIA/Watchers	Local
Local Storage	/Volumes/DAS_VOLUME	Local

If you elected to have an add-only scan done, then after the prescribed amount of time, the Final Cut Server should begin to access the device and start creating assets, proxy files, and so on. But don't hover over the machine and wait for it to start—that's like watching a pot of water boil. Go get yourself a nice beverage and be surprised when you come back to find it humming along. Depending on the number of files that need to be cataloged, this process may take some time before the assets show up in the client application.

Creating Simple Automations

Now that users can log in, and assets are beginning to show up in the client application, you can take advantage of all the distracted oohing and aahing to knuckle down and design the true gems of your already thankless job: automations. Although the choices are quite limited within the confines of System Preferences, this is a good way to first experience the power of a well-constructed automation. When you're ready to tweak to your heart's content and have full access to everything that Final Cut Server can automate, Lesson 8 will deliver. For right now, let's dig in and see what we can get done in the GUI of the preference pane.

1 Click the Automations tab to open the Automations pane.

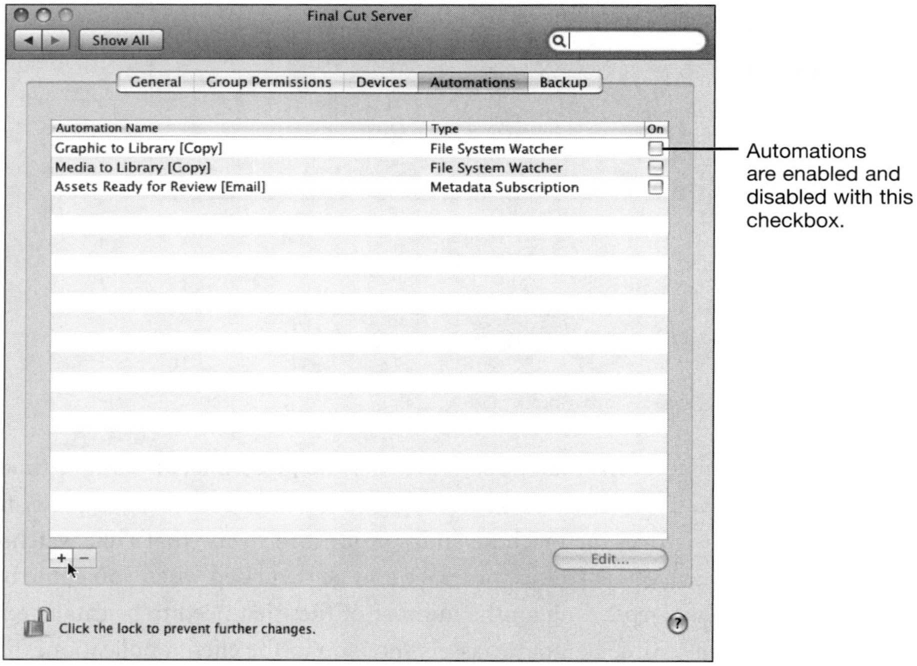

2 Click the plus (+) button to launch the Automation Setup Assistant.

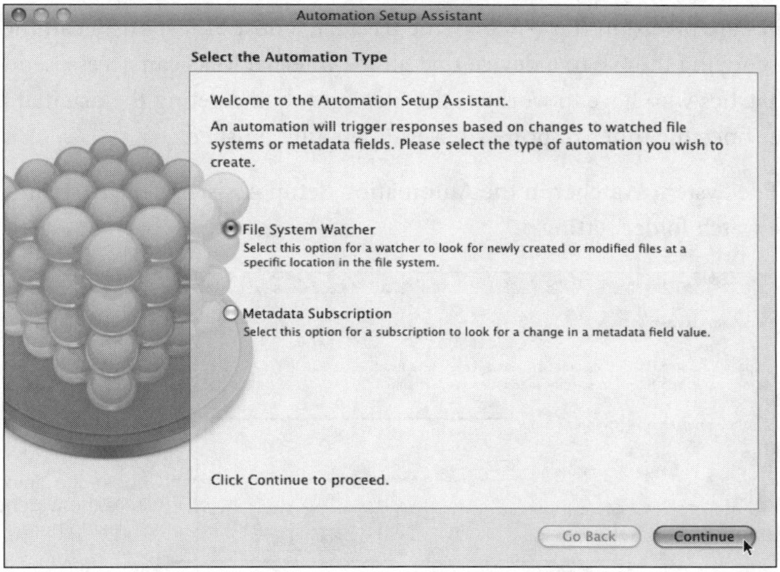
Automation Setup Assistant
Select the Automation Type
Welcome to the Automation Setup Assistant.
An automation will trigger responses based on changes to watched file systems or metadata fields. Please select the type of automation you wish to create.
File System Watcher
Select this option for a watcher to look for newly created or modified files at a specific location in the file system.
Metadata Subscription
Select this option for a subscription to look for a change in a metadata field value.
Click Continue to proceed.
Go Back
Continue

The Automation Setup Assistant offers two types of automation, file system watcher and metadata subscription—a limited selection, as we said, but powerful. Instructions for creating each kind of automation are explained in the next two sections.

Defining File System Watchers

Watchers are a practical gateway to the devices of your organization. They ensure that files that are dropped inside them are automatically transcoded and copied to the appropriate devices, and that assets are created in the catalog, so that everyone can search, view, annotate, and use them in their work.

The event of a file being placed within a watcher is called a trigger, one of many that you can designate in Final Cut Server. The actions that Final Cut Server performs as a result of a trigger are called responses.

Here's an example of a watcher in action. Say you have folks outside of your organization who need to contribute content gathered in the field. You can create a watcher within an FTP site. When files are placed in the watcher (the trigger), a host of responses can then fire off, including copying the file to a device that all the internal folks can access, sending emails to the parties who have to work on the files, and then deleting the original files within the watcher once the other responses have successfully finished.

When you select File System Watcher in the Automation Setup Assistant, the first pane asks you to define watch folder settings.

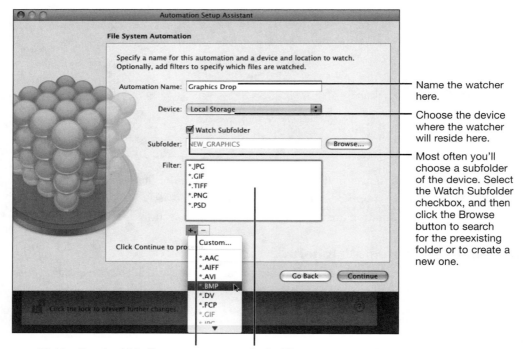

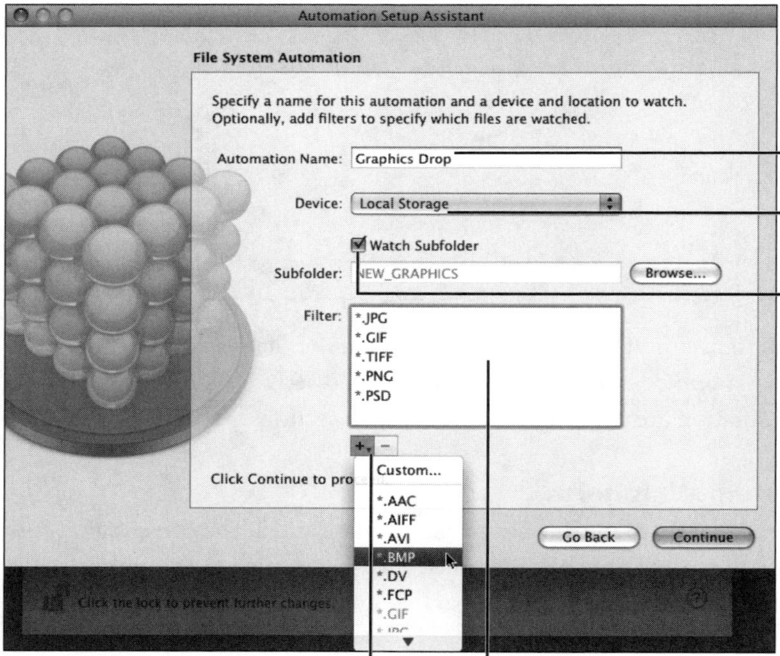

Name the watcher here.

Choose the device where the watcher will reside here.

Most often you'll choose a subfolder of the device. Select the Watch Subfolder checkbox, and then click the Browse button to search for the preexisting folder or to create a new one.

Clicking the plus (+) button opens a pop-up menu that allows you to choose from a series of predefined filename-extension filters. The Custom selection at the top of the list allows you to define any kind of filename filter.

In the Filter area, you can instruct the watcher to look for specific filename extensions. In this case, the filter will find only image files.

Click Continue to define the responses for the watcher. Then, click the plus (+) button at lower left to choose your responses.

The Copy response is key in a watcher, because it transports the file to another device, perhaps transcoding it along the way.

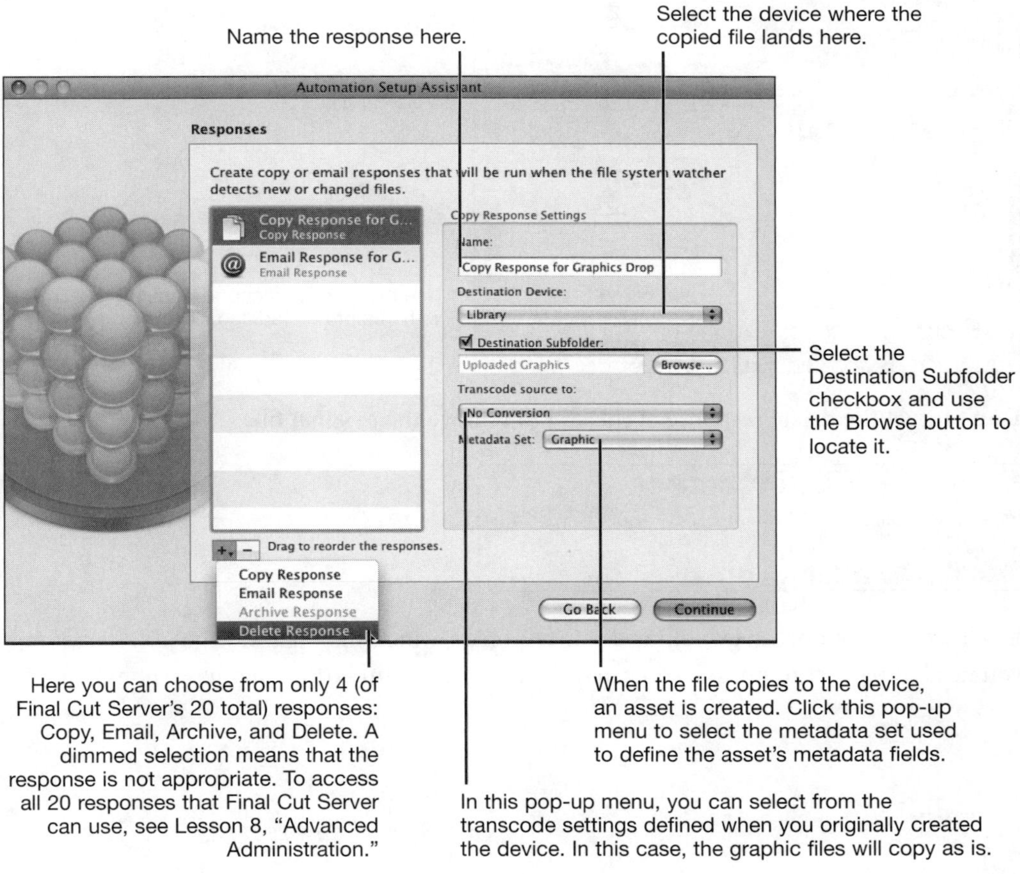
Name the response here.
Select the device where the copied file lands here.
Automation Setup Assistant
Responses
Create copy or email responses that will be run when the file system watcher detects new or changed files.
Copy Response for G...
Copy Response
Email Response for G...
Email Response
Copy Response Settings
Name:
Copy Response for Graphics Drop
Destination Device:
Library
Destination Subfolder:
Uploaded Graphics
Browse...
Transcode source to:
No Conversion
Metadata Set:
Graphic
Drag to reorder the responses.
Copy Response
Email Response
Archive Response
Delete Response
Go Back
Continue
Select the Destination Subfolder checkbox and use the Browse button to locate it.
Here you can choose from only 4 (of Final Cut Server's 20 total) responses: Copy, Email, Archive, and Delete. A dimmed selection means that the response is not appropriate. To access all 20 responses that Final Cut Server can use, see Lesson 8, "Advanced Administration."
When the file copies to the device, an asset is created. Click this pop-up menu to select the metadata set used to define the asset's metadata fields.
In this pop-up menu, you can select from the transcode settings defined when you originally created the device. In this case, the graphic files will copy as is.

Email responses give us the ability to contact certain people to inform them of the arrival of files, or of the progression of a project to the next level. The coolest part of the response is the ability to use metadata fields in the email so that it will be customized for the circumstance.

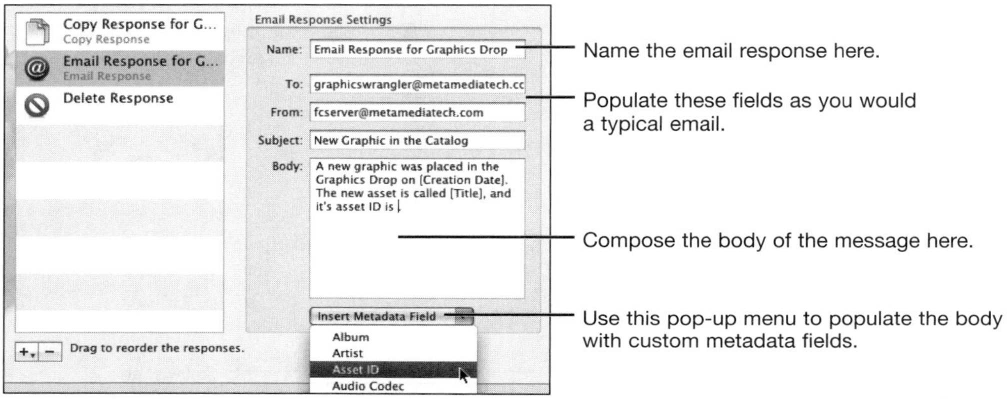
Copy Response for G...
Copy Response
Email Response for G...
Email Response
Delete Response
Drag to reorder the responses.
Email Response Settings
Name: Email Response for Graphics Drop
To: graphicswrangler@metamediatech.cc
From: fcserver@metamediatech.com
Subject: New Graphic in the Catalog
Body: A new graphic was placed in the Graphics Drop on [Creation Date]. The new asset is called [Title], and it's asset ID is
Insert Metadata Field
Album
Artist
Asset ID
Audio Codec
Name the email response here.
Populate these fields as you would a typical email.
Compose the body of the message here.
Use this pop-up menu to populate the body with custom metadata fields.

Don't forget the delete response at the end to remove the original file.

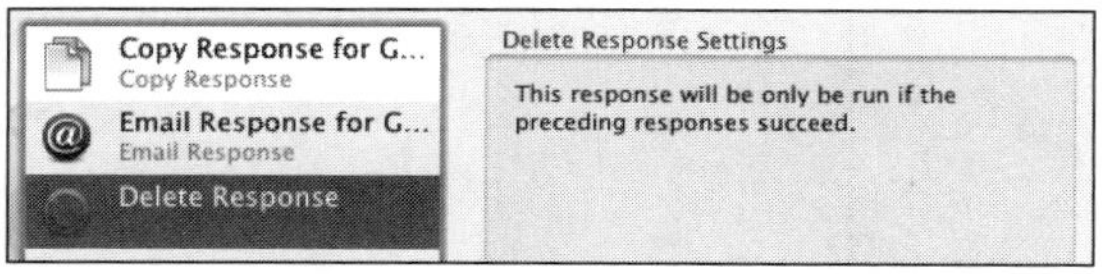
Copy Response for G...
Copy Response
Email Response for G...
Email Response
Delete Response
Delete Response Settings
This response will be only be run if the preceding responses succeed.

Don't worry about order, except for delete responses, which should always go last. In case you need to reorder responses, you can do so by dragging and dropping the responses at left.

When you click the Continue button at lower left, a summary pane appears.

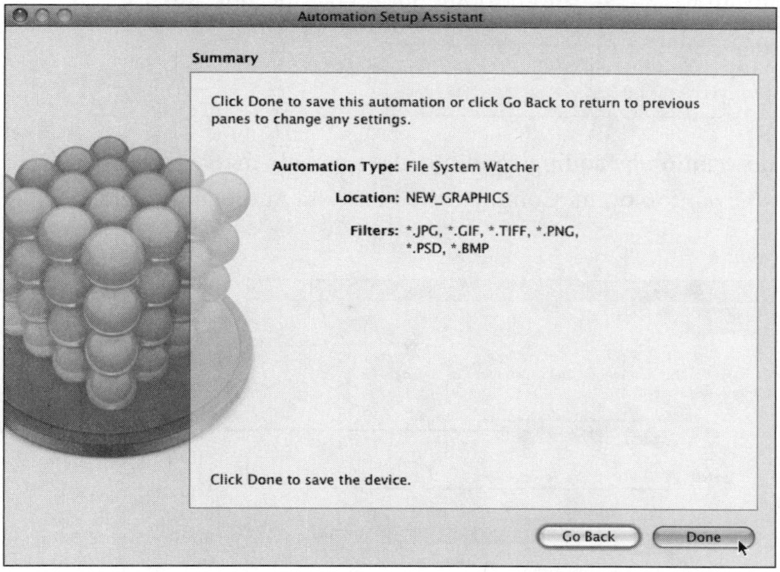

Click Done to save the automation and immediately start it.

Back in the main window, the automation now appears, enabled and ready to roll.

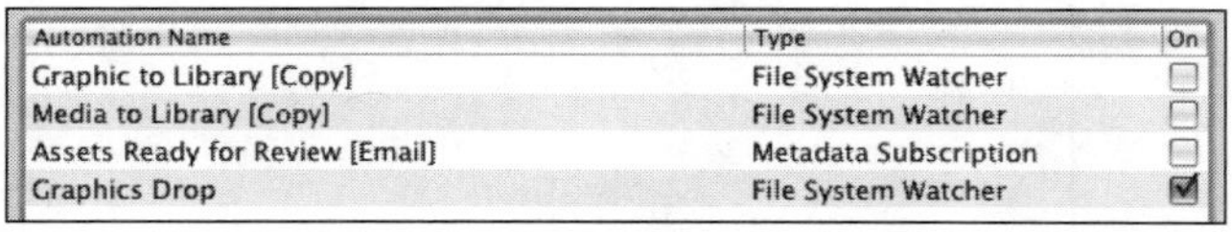

Automation Name	Type	On
Graphic to Library [Copy]	File System Watcher	☐
Media to Library [Copy]	File System Watcher	☐
Assets Ready for Review [Email]	Metadata Subscription	☐
Graphics Drop	File System Watcher	☑

To test the watcher, plop some files in the folder, wait a few moments, and then check both the catalog and the folder.

Defining Metadata Subscriptions

Metadata subscriptions trigger responses that occur when specific metadata has been changed, usually by your end users.

In this example, you'll automatically archive an asset if anyone with appropriate access changes the asset's status to Completed.

Back in the Automations pane, click the plus (+) button to launch the Automation Setup Assistant. This time, click the Metadata Subscription button before continuing.

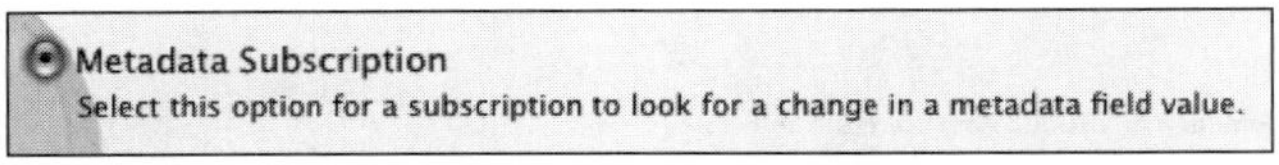
Metadata Subscription
Select this option for a subscription to look for a change in a metadata field value.

Define the metadata subscription by adding conditional statements that, when combined, shape the trigger that you want to occur. Complete the Metadata Automation fields as shown below.

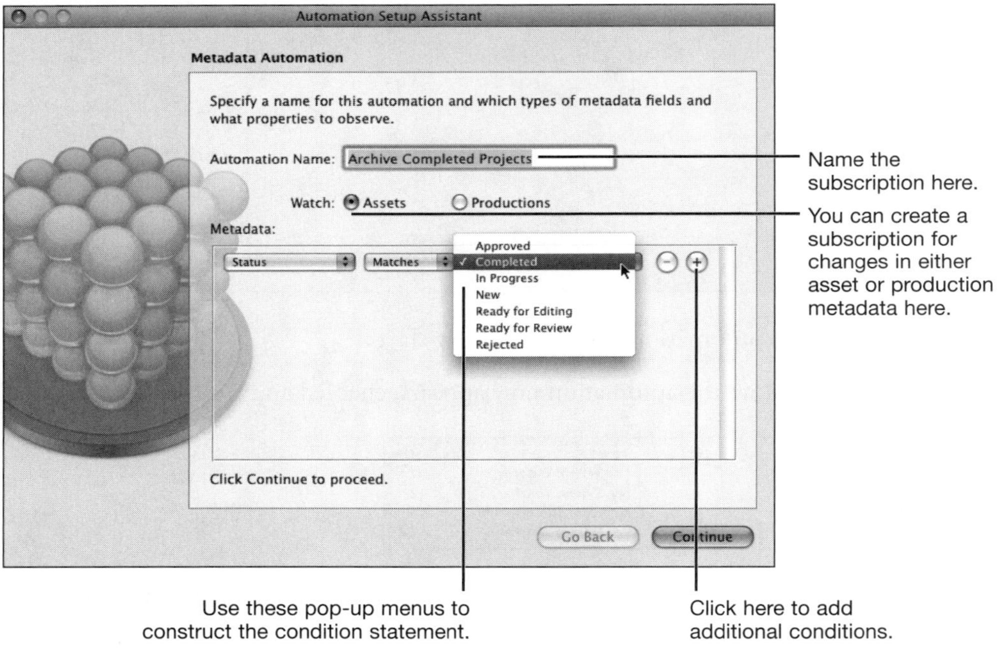
Automation Setup Assistant
Metadata Automation
Specify a name for this automation and which types of metadata fields and what properties to observe.
Automation Name: Archive Completed Projects
Watch: Assets
Productions
Metadata:
Status
Matches
Approved
Completed
In Progress
New
Ready for Editing
Ready for Review
Rejected
Click Continue to proceed.
Go Back
Continue
Name the subscription here.
You can create a subscription for changes in either asset or production metadata here.
Use these pop-up menus to construct the condition statement.
Click here to add additional conditions.

The Responses pane is exactly the same as it is in the watcher assistant.

Here's what an archive response might look like.

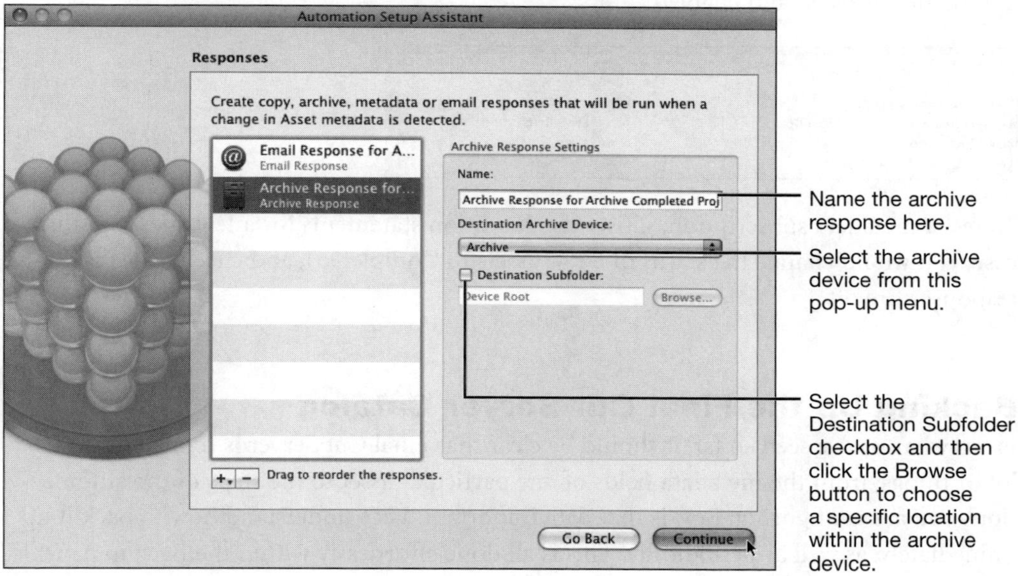
Automation Setup Assistant
Responses
Create copy, archive, metadata or email responses that will be run when a change in Asset metadata is detected.
Email Response for A...
Email Response
Archive Response for...
Archive Response
Archive Response Settings
Name:
Archive Response for Archive Completed Proj
Destination Archive Device:
Archive
Destination Subfolder:
Device Root
Browse...
Drag to reorder the responses.
Go Back
Continue
Name the archive response here.
Select the archive device from this pop-up menu.
Select the Destination Subfolder checkbox and then click the Browse button to choose a specific location within the archive device.

When you click Continue, a summary pane appears.

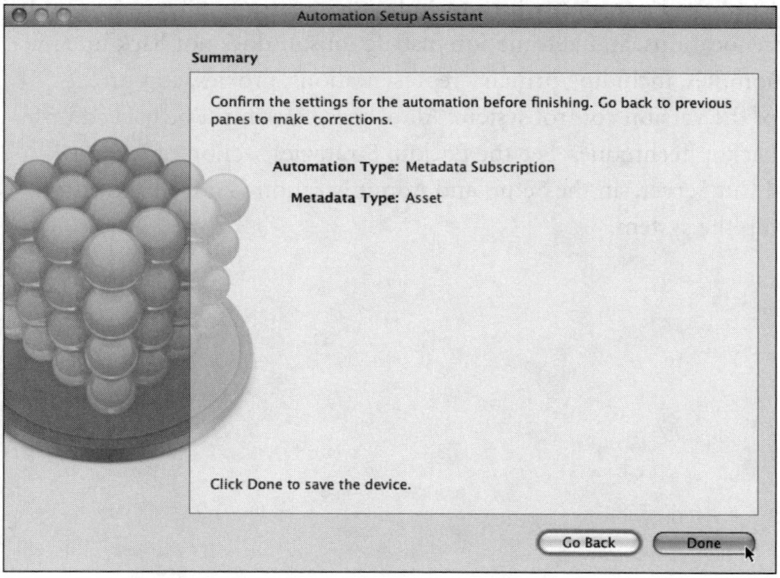
Automation Setup Assistant
Summary
Confirm the settings for the automation before finishing. Go back to previous panes to make corrections.
Automation Type: Metadata Subscription
Metadata Type: Asset
Click Done to save the device.
Go Back
Done

Click Done to finish the automation. When you return to the Automations pane, the automation is listed and enabled.

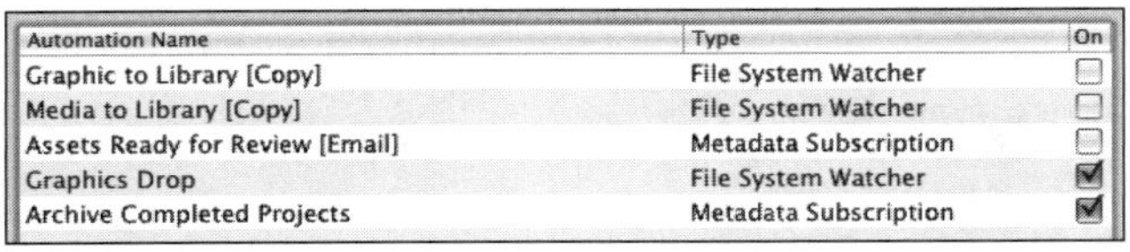

To test a metadata subscription, satisfy the condition statements for a few assets (in this case, you would change the status of a few assets to Completed), and check to see if the responses fire off.

Backing Up the Final Cut Server Catalog

From what you've seen so far, it should be clear that Final Cut Server is remembering a lot of things, from the metadata fields of any particular asset to the steps of the automations you are configuring. Boy, is that data important. Let's understand how to back it up immediately, as well as periodically. This is all done effortlessly within the Backup pane.

> **NOTE ►** It's crucial to understand that the Backup pane accomplishes the task of backing up Final Cut Server's catalog, which includes the records of all its assets and productions, device locations, and automation instructions. It does not back up any of the representation files, including primary representations, proxies, edit proxies, and the contents of the version control system. All of these files must be backed up using traditional backup techniques. See the Backup Strategies section of Lesson 7, "Backing Up Final Cut Server," in the Setup and Administration Guide, for ideas on how to fully back up the system.

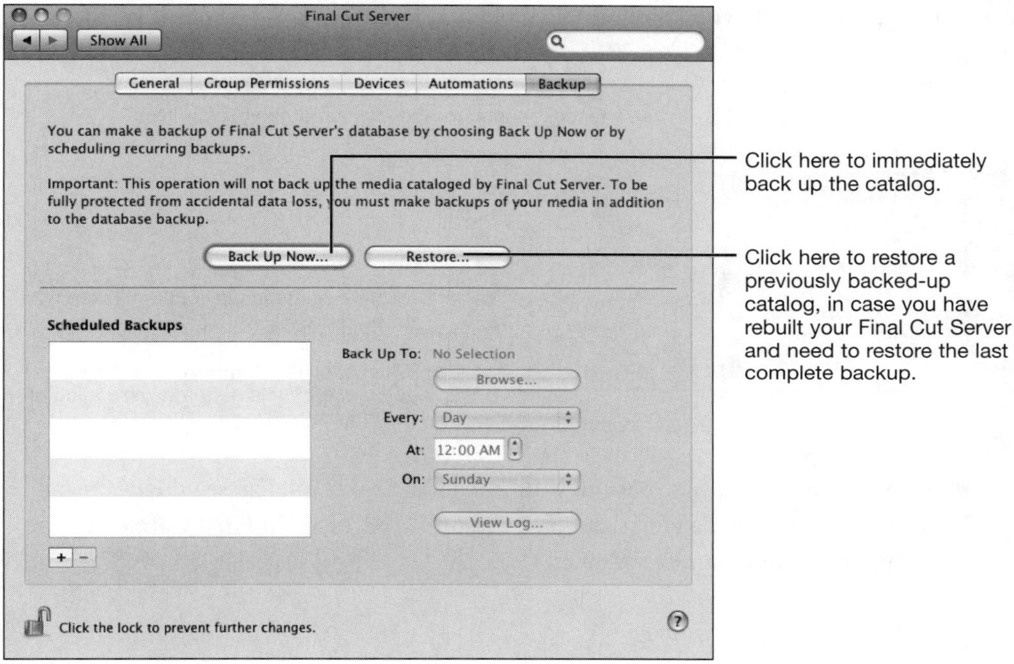
Final Cut Server
Show All
General
Group Permissions
Devices
Automations
Backup
You can make a backup of Final Cut Server's database by choosing Back Up Now or by scheduling recurring backups.
Important: This operation will not back up the media cataloged by Final Cut Server. To be fully protected from accidental data loss, you must make backups of your media in addition to the database backup.
Back Up Now...
Restore...
Scheduled Backups
Back Up To: No Selection
Browse...
Every: Day
At: 12:00 AM
On: Sunday
View Log...
Click the lock to prevent further changes.
Click here to immediately back up the catalog.
Click here to restore a previously backed-up catalog, in case you have rebuilt your Final Cut Server and need to restore the last complete backup.

Click Back Up Now to bring up a simple dialog.

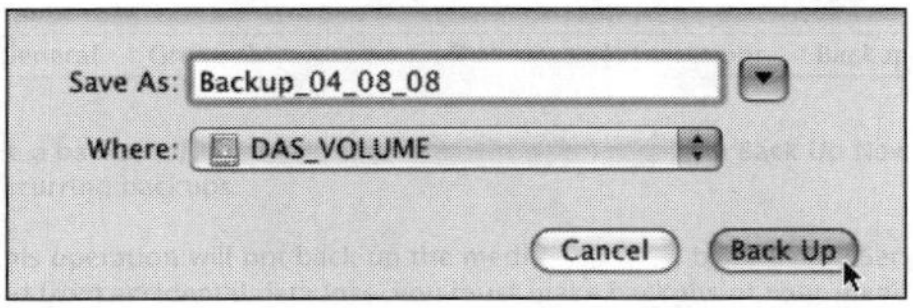
Save As: Backup_04_08_08
Where: DAS_VOLUME
Cancel
Back Up

Select a location and a filename and click Back Up. Remember to use a location that is *not* on the same storage as your Final Cut Server software.

To create a scheduled backup, click the plus (+) button at the lower left of the Backup pane.

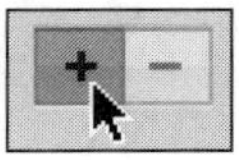

An entry appears at left, with configurable settings at right.

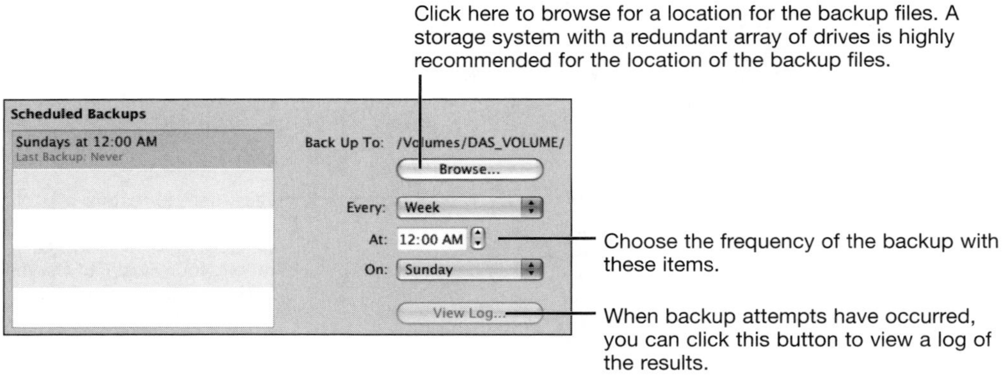
Click here to browse for a location for the backup files. A storage system with a redundant array of drives is highly recommended for the location of the backup files.
Scheduled Backups
Sundays at 12:00 AM
Last Backup: Never
Back Up To: /Volumes/DAS_VOLUME/
Browse...
Every: Week
At: 12:00 AM
On: Sunday
View Log...
Choose the frequency of the backup with these items.
When backup attempts have occurred, you can click this button to view a log of the results.

NOTE ► If you ever have to restore your Final Cut Server catalog, see the Restoring Your Final Cut Server Catalog section of Lesson 7, "Backing Up Final Cut Server," in the Setup and Administration Guide.

Lesson Review

1. Once you have created groups in Accounts in System Preferences, how would you add these groups to Final Cut Server?
2. What are permission sets and how do they affect a user's interaction with Final Cut Server?
3. What is the default number of jobs that Final Cut Server processes?
4. If a user is a member of more than one group, and these groups have different permission sets, how does Final Cut Server give that user access?
5. How can the administrator add and edit devices within Final Cut Server?
6. What is the difference between a full scan and an add only scan?

Answers

1. Using the Final Cut Server System Preferences pane, click on the Group Permissions tab, then click the plus button to add the group.
2. Permissions Sets are a combination of functionality (what a user can do) and access (what assets and productions a user can see) settings. When the group a user belongs to is assigned a permission set, then all users will have the functionality and access defined within that permission set.
3. Final Cut Server processes ten jobs by default, but the administrator can alter the number.
4. The administrator can set the priority levels of the user's groups so that the group with the highest level determines the permission set used during log in.
5. Use the Device Setup Assistant in the Final Cut Server System Preferences pane to add and edit devices.
6. A full scan examines the device for any content changes, including any new, changed, or removed files. These changes are reflected in the Final Cut Server catalog. Full scans are processor intensive and should be done at a time when your organization isn't busy. An Add Only scan adds only new assets based on newly added files. This type of scan is usually configured to occur more frequently.

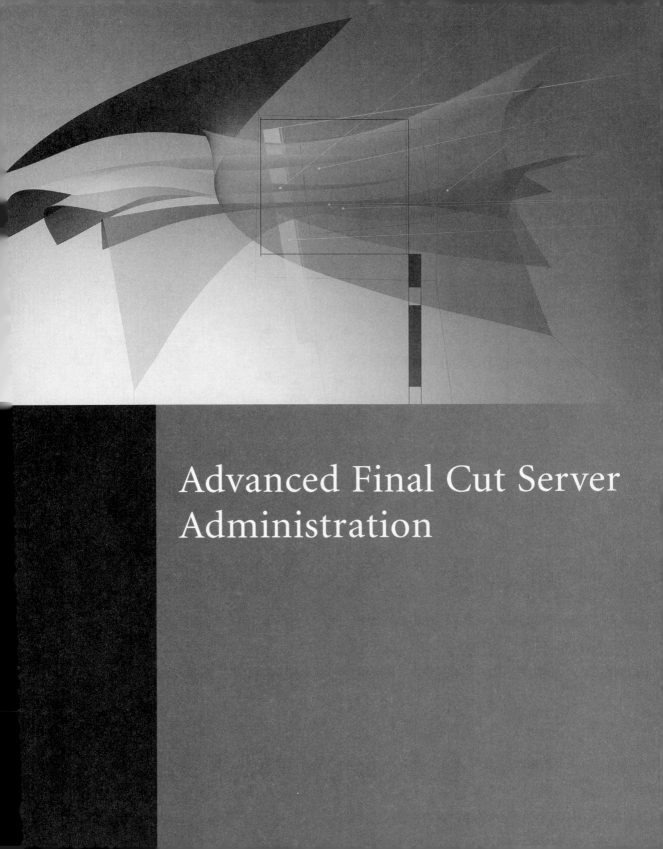

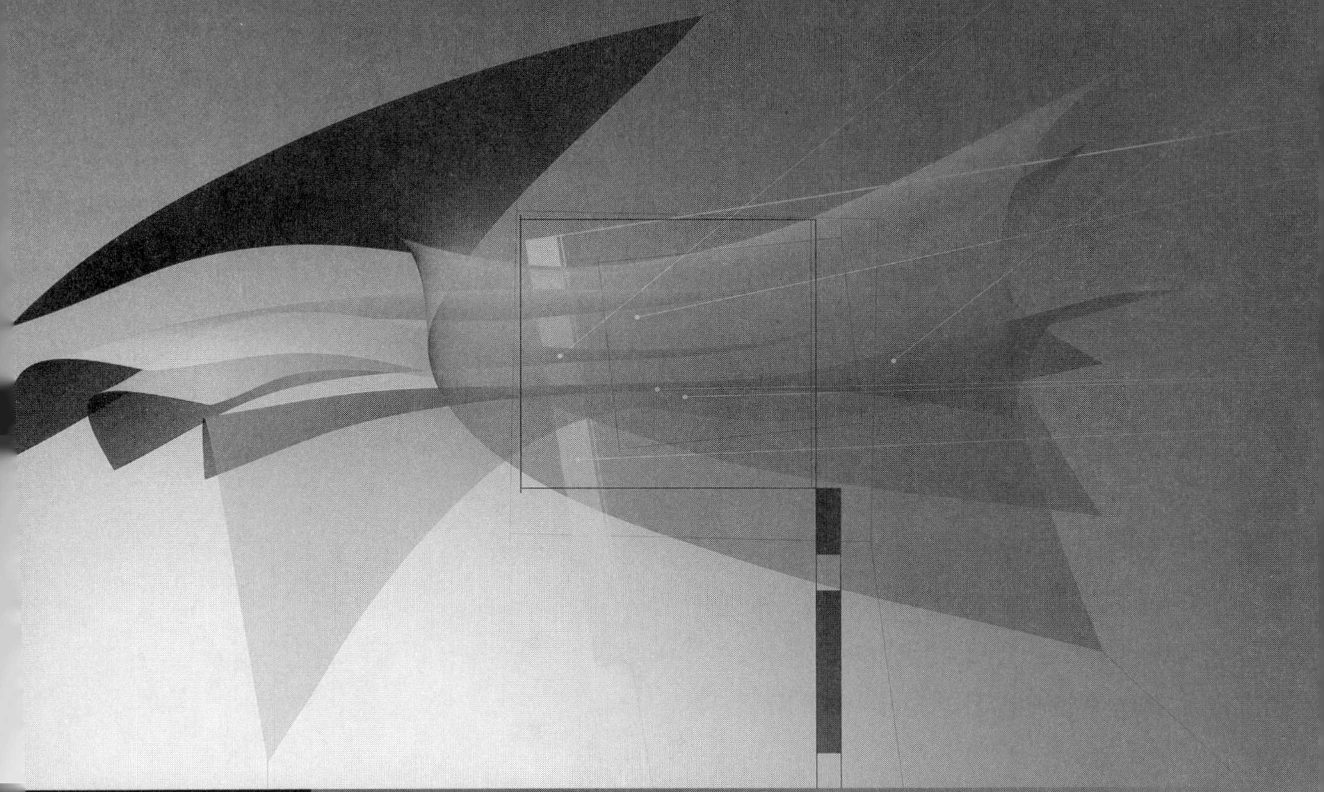

Advanced Final Cut Server Administration

8

Goals

Add and edit metadata structures
Add edit-in-place status to a device
Create customized transcode settings
Create advanced automations
Add and edit permission sets
Configure Final Cut Server preferences

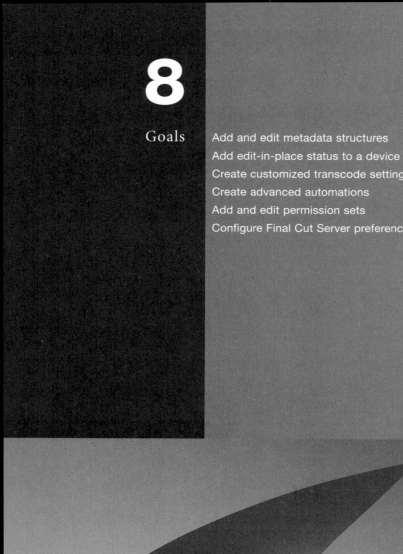

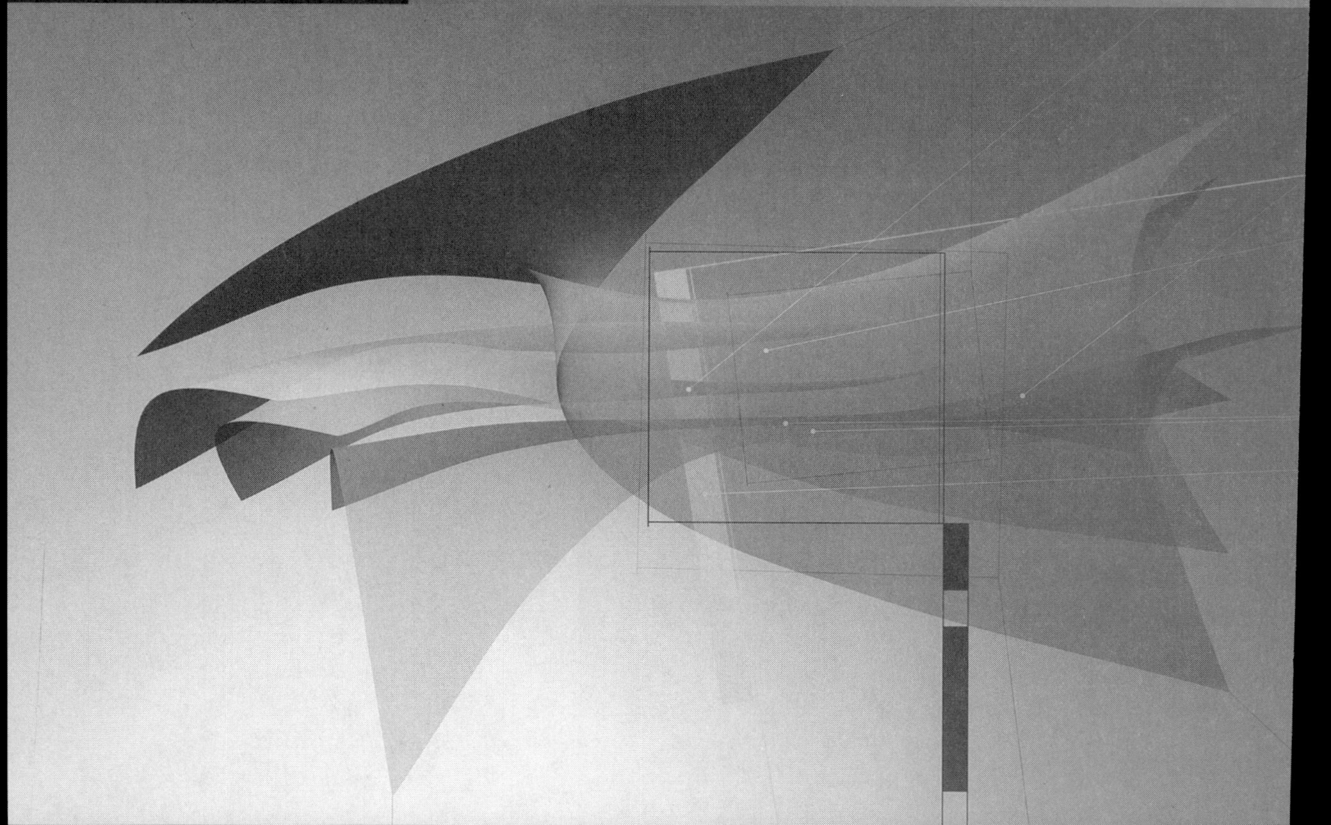

Lesson **8**

Advanced Administration

This lesson covers just about everything that's left to know as an administrator of a Final Cut Server, including the more granular control that you've probably been yearning for since reading Lesson 7. All tasks will be accomplished within the Administration Panel of the client application. This window is only accessible to users who are logged in with admin privileges.

With greater control comes greater complexity, so it's best to caffeinate heavily before diving into these pages. Many tasks are accomplished with the end in mind, which actually means constructing things backwards. However, once you've internalized the process with a few practice runs, you'll be automating and customizing your organization's Final Cut Server like a maven.

> **MORE INFO ▸** Brevity is a necessity in a Getting Started book. The Setup and Administration Guide that came with your software offers more detailed descriptions of the functions of some of the tools in the Administration Panel. Start here first to get the big picture, and refer to that source if you need more detail.

Administration Tools Within the Client Application

The Administration Panel is accessible to users who belong to the admin permission set. The admin user of the Final Cut Server is always a member of this group.

Select Administration from the Server pop-up menu in the upper left of the main Client Application window to open the Administration Panel.

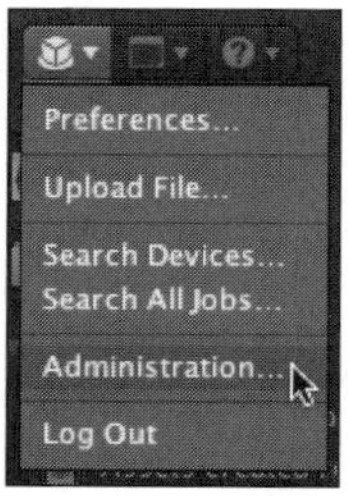
Preferences...
Upload File...
Search Devices...
Search All Jobs...
Administration...
Log Out

Before displaying the window, Final Cut Server warns that you're headed into serious territory.

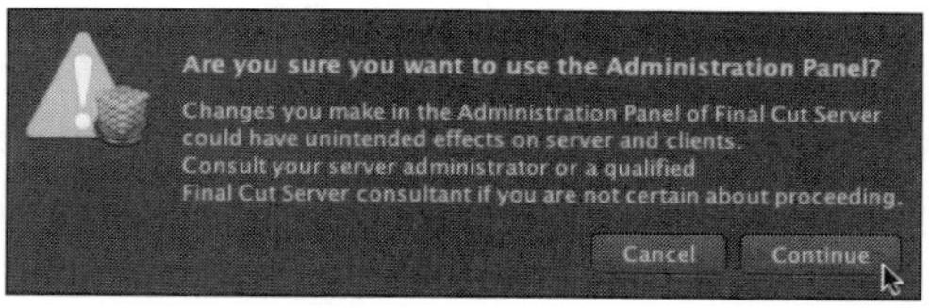
Are you sure you want to use the Administration Panel?
Changes you make in the Administration Panel of Final Cut Server could have unintended effects on server and clients.
Consult your server administrator or a qualified Final Cut Server consultant if you are not certain about proceeding.
Cancel
Continue

At the left side of the window, you'll see tabs for different tasks. This area is the heart of the Administration Panel. You'll see these tabs at the beginning of each section to show you where to click to accomplish a task. We'll be working out of order from the list you see, going to the most accessed tasks first.

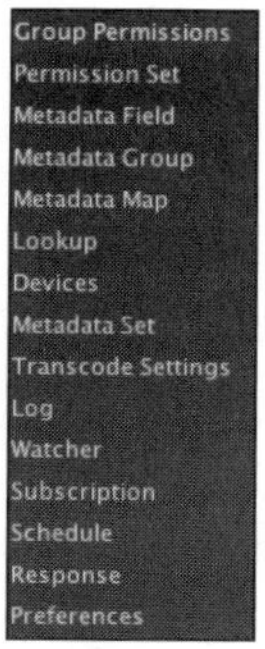
Group Permissions
Permission Set
Metadata Field
Metadata Group
Metadata Map
Lookup
Devices
Metadata Set
Transcode Settings
Log
Watcher
Subscription
Schedule
Response
Preferences

Tabs in the Administration Panel

Adding and Editing Metadata

Before you dive in to adding and editing metadata, it's important to understand how Final Cut Server organizes its metadata.

Every piece of metadata is stored in a metadata field, which are then organized into metadata groups. Several metadata groups form a metadata set. Metadata sets are the largest building block of metadata and are used to define an asset or production.

Sometimes you'll want to restrict the values that a user can enter into a metadata field, so instead of giving them a blank box to fill in, you'll give them a pop-up menu instead. Final Cut Server calls these kinds of choice-based systems *lookups*.

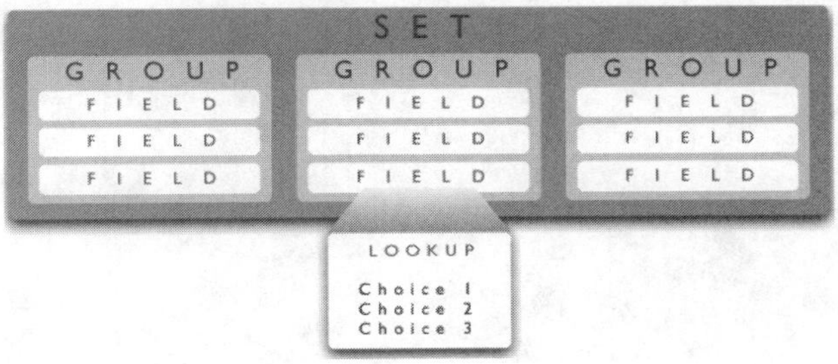

To put these into practical perspective, when looking at an asset's info window, fields show up in the Metadata tab. The list of items at the left is the metadata groups, and the "definition" of the asset is the metadata set. Fields that have pop-up menus have a lookup assigned to them.

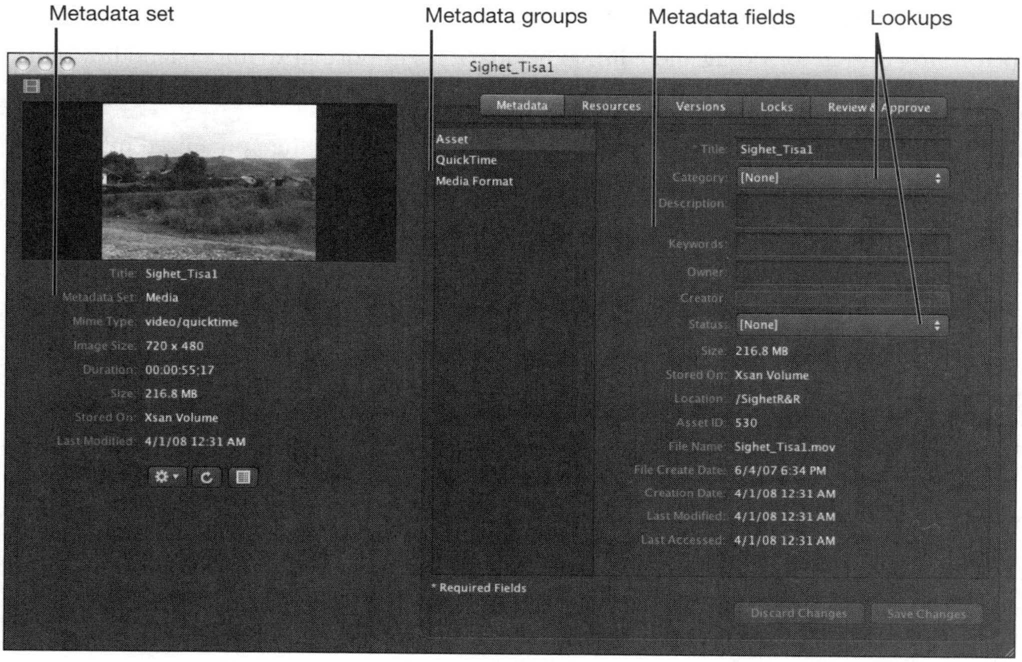

As you can see, the column for the metadata groups is just *waiting* to be filled with customized groups, populated with customized fields—all specialized for your organization's workflow. To illustrate how it's done, the following examples will create a metadata group that will hold important information about the client associated with the asset: their name, job title, job number, and lead contact email address. Each of these four pieces of information will have an appropriate field created for it. We'll then create a client information metadata group to collect these four fields. While doing so, a lookup will be assigned for the client's name field to make sure proper searching and automations can be accomplished. Finally, we'll add this new group to the appropriate metadata sets so this group will appear for all of our assets and productions.

Metadata groups have an additional function within Final Cut Server: They determine what fields are used in advanced searches, subscriptions, and list views. These too are customizable. Later in this section, we'll see how to modify some custom metadata groups that govern the fields used in filtering for advanced searches and metadata subscriptions.

We'll also change a list view metadata group that dictates what fields are shown when viewing the results of a search.

Lastly, we'll cover the powerful Metadata Map pane, and how it lets us shuttle metadata in and out of files that are entering and exiting Final Cut Server.

We'll do all of this by visiting the metadata tabs found in the Administration Panel.

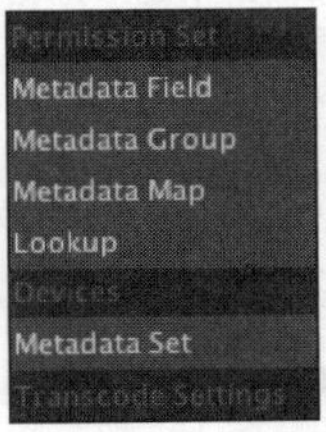

The Metadata tabs

Once you click on their respective tabs, all of the metadata-based panes in the Administration Panel look the same. There is a search area and a single Create New Item button. Double-clicking on existing items allows you to view and edit them.

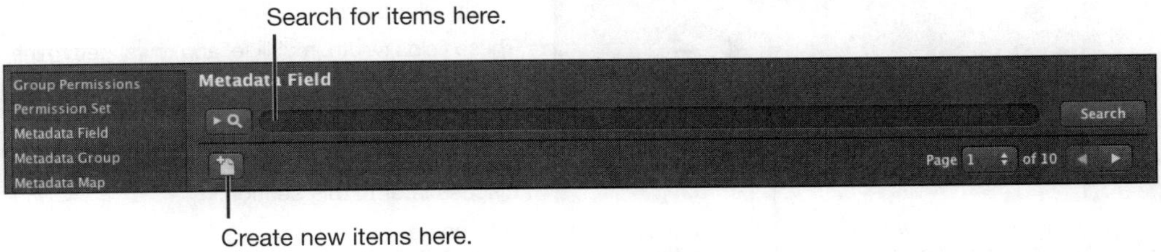

There is also a shortcut menu that you can access by right-clicking on an existing item within the pane.

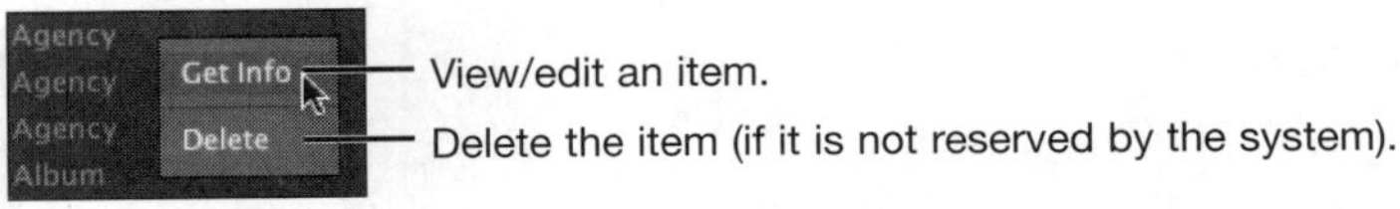

Creating and Editing Metadata Fields

Start the customization process by first creating the new metadata fields that will hold your custom metadata. This is accomplished with the Metadata Field pane.

In this example, we are creating a text-based field to describe the client's name.

1 Click the Metadata Field tab, at left, to open the Metadata Field pane.

2 Click the Add Item button to reveal the Metadata Field window.

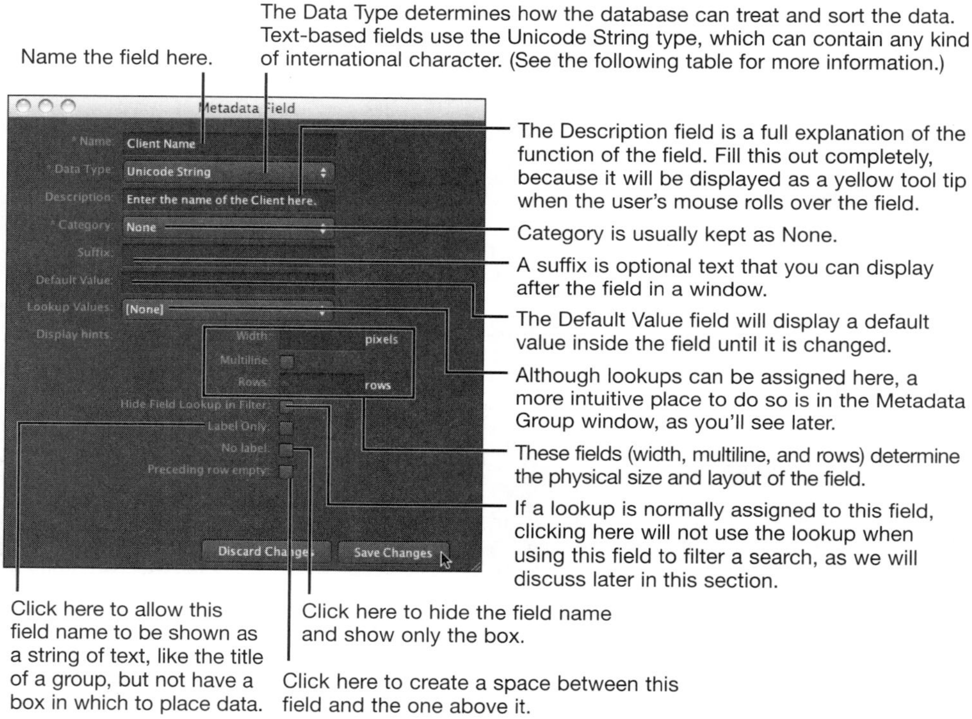
The Data Type determines how the database can treat and sort the data. Text-based fields use the Unicode String type, which can contain any kind of international character. (See the following table for more information.)
Name the field here.
Metadata Field
* Name: Client Name
* Data Type: Unicode String
Description: Enter the name of the Client here.
* Category: None
Suffix:
Default Value:
Lookup Values: [None]
Display hints:
Width: pixels
Multiline:
Rows: rows
Hide Field Lookup in Filter:
Label Only:
No label:
Preceding row empty:
Discard Changes
Save Changes
The Description field is a full explanation of the function of the field. Fill this out completely, because it will be displayed as a yellow tool tip when the user's mouse rolls over the field.
Category is usually kept as None.
A suffix is optional text that you can display after the field in a window.
The Default Value field will display a default value inside the field until it is changed.
Although lookups can be assigned here, a more intuitive place to do so is in the Metadata Group window, as you'll see later.
These fields (width, multiline, and rows) determine the physical size and layout of the field.
If a lookup is normally assigned to this field, clicking here will not use the lookup when using this field to filter a search, as we will discuss later in this section.
Click here to allow this field name to be shown as a string of text, like the title of a group, but not have a box in which to place data.
Click here to hide the field name and show only the box.
Click here to create a space between this field and the one above it.

3 Click Save Changes to save the field.

The Data Type in this example is Unicode String. There are six other Data Types and the table on the following page explains their differences.

Selecting different Data Types will reveal different Display Hints, listed at the bottom of the window. See the Setup and Administration Guide for descriptions of what these particular settings accomplish. In the example that follows, we can see different display hints when creating the client job number metadata field with an Integer Data Type.

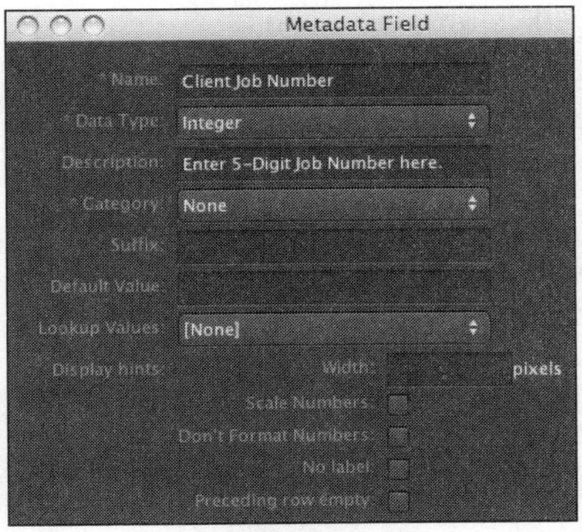

Data Types

Data Type	Description
Boolean	True/false or yes/no state
Date	Date and time of day
Floating Point Number	Number that needs a decimal place
Integer	Number that doesn't need a decimal place, up to 32 bits in size
Large Integer	64-bit number, just in case you need it
Timecode	Timecode stamp, using the format hh:mm:ss:ff
Unicode String	Generic text field, using any Unicode character

4 Repeat steps 1 through 3 to create three other fields: Client Job Title, Client Contact Email (using Unicode String data type for both), and Client Job Number (using Integer data type, as seen in the picture above).

Creating and Editing Lookups for Metadata Fields

Lookups are pop-up menus that pre-determine the end user's choices in a metadata field.

In the following example, we will create a lookup for the Client Name metadata field (created in the previous section) that will force users to choose from the organization's current client list. One by one, the options for the lookup are added. More items can be added to the lookup by editing it (so if you get more clients, you can add them to the list later).

1 Click the Lookup tab, at left, to open the Lookup pane.

2 Click the Add Item button to reveal the Lookup window.

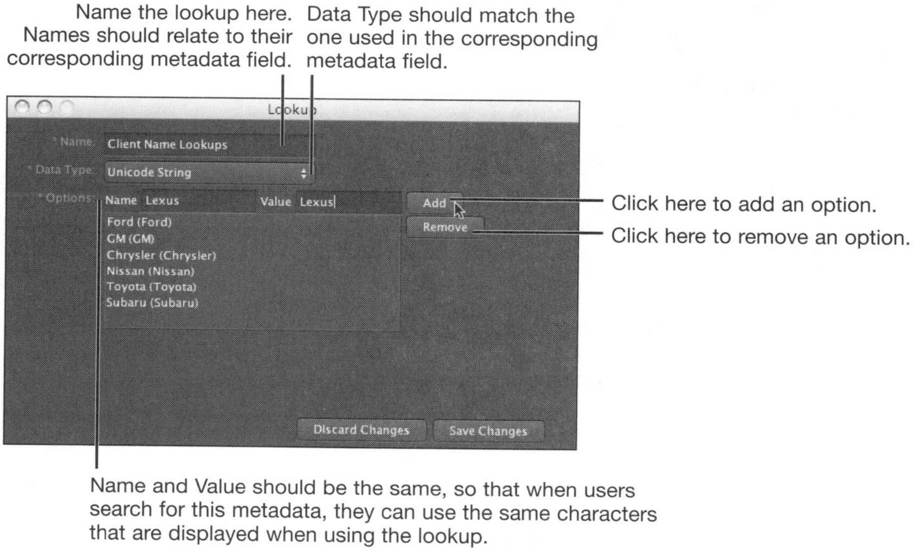
Name the lookup here. Names should relate to their corresponding metadata field.
Data Type should match the one used in the corresponding metadata field.
Lookup
Name:
Client Name Lookups
Data Type:
Unicode String
Options:
Name Lexus
Value Lexus
Add
Remove
Ford (Ford)
GM (GM)
Chrysler (Chrysler)
Nissan (Nissan)
Toyota (Toyota)
Subaru (Subaru)
Discard Changes
Save Changes
Click here to add an option.
Click here to remove an option.
Name and Value should be the same, so that when users search for this metadata, they can use the same characters that are displayed when using the lookup.

3 Click Save Changes to save the lookup.

NOTE ▶ There is no association between this lookup and its corresponding metadata field as yet. This will be done in the next section when we work with metadata groups.

Creating and Editing Metadata Groups

Now that we've created all the new fields and lookups, the Metadata Group pane is where everything comes together. In one window, the four fields are organized into a Client Information group, the Client Name lookup is assigned to the Client Name field, and the group itself is associated with one or more metadata sets, such as Media (for assets) and Show (for productions).

1 Click the Metadata Group tab, at left, to open the Metadata Group pane.

2 Click the Add Item button to reveal the Metadata Group window.

The Metadata Groups window is quite long, so below it's broken down into two parts.

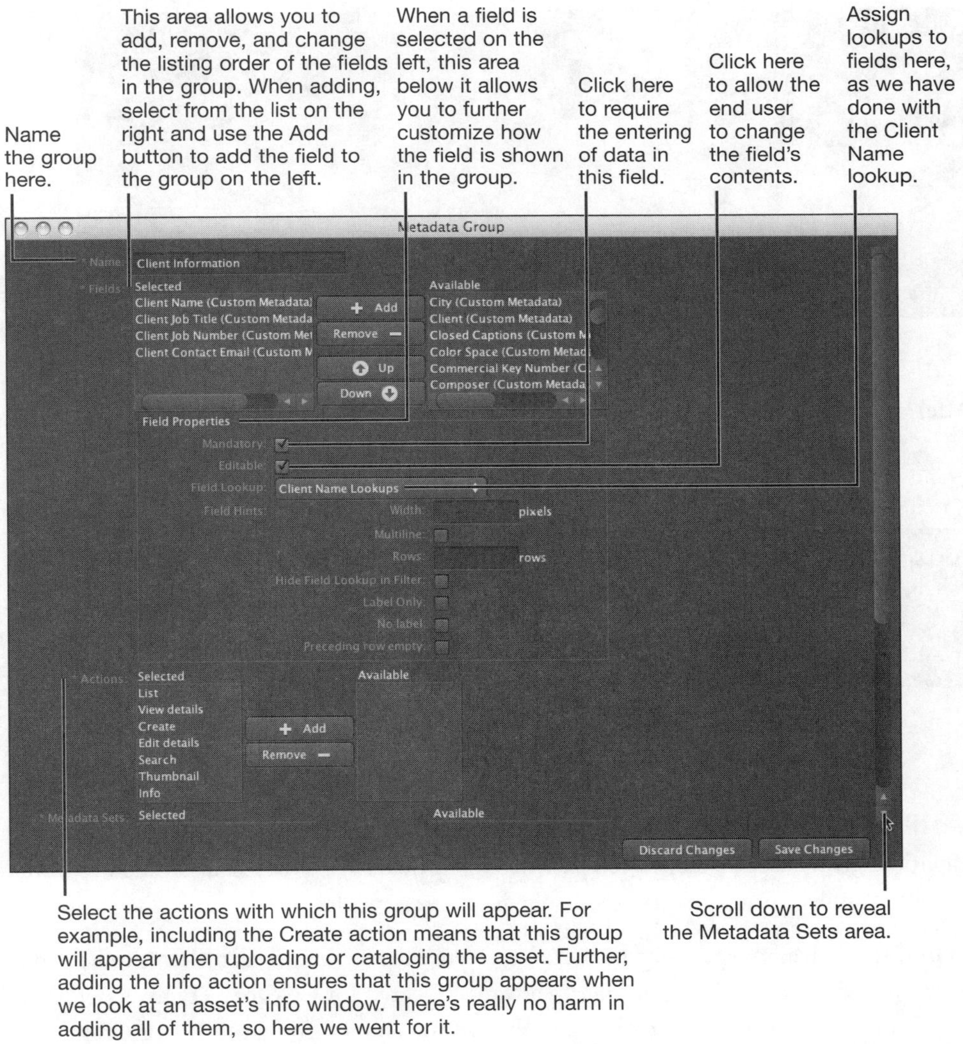
Name the group here.
This area allows you to add, remove, and change the listing order of the fields in the group. When adding, select from the list on the right and use the Add button to add the field to the group on the left.
When a field is selected on the left, this area below it allows you to further customize how the field is shown in the group.
Click here to require the entering of data in this field.
Click here to allow the end user to change the field's contents.
Assign lookups to fields here, as we have done with the Client Name lookup.
Metadata Group
Name
Client Information
Fields
Selected
Client Name (Custom Metadata)
Client Job Title (Custom Metada
Client Job Number (Custom Met
Client Contact Email (Custom M
Add
Remove
Up
Down
Available
City (Custom Metadata)
Client (Custom Metadata)
Closed Captions (Custom M
Color Space (Custom Metad
Commercial Key Number (C
Composer (Custom Metada
Field Properties
Mandatory
Editable
Field Lookup
Client Name Lookups
Field Hints
Width
pixels
Multiline
Rows
rows
Hide Field Lookup in Filter
Label Only
No label
Preceding row empty
Actions
Selected
List
View details
Create
Edit details
Search
Thumbnail
Info
Add
Remove
Available
Metadata Sets
Selected
Available
Discard Changes
Save Changes
Select the actions with which this group will appear. For example, including the Create action means that this group will appear when uploading or cataloging the asset. Further, adding the Info action ensures that this group appears when we look at an asset's info window. There's really no harm in adding all of them, so here we went for it.
Scroll down to reveal the Metadata Sets area.

The bottom of the window is where you assign this group to one or more preexisting metadata sets.

This area allows you to associate this group with metadata sets. In this case, since we want this new group to appear everywhere in the catalog, we've added all the current sets. If you're creating a new metadata set in which to place this group, leave this area blank, save your changes, and proceed to the next section on metadata sets.

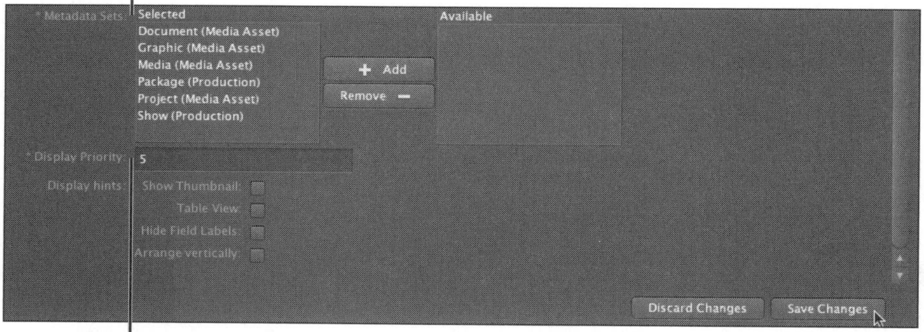

Control the listing order groups within a set here.

After saving your changes, you and your users can start using the new group immediately.

The asterisk denotes the mandatory nature of this field, as we specified within the Metadata Group window.

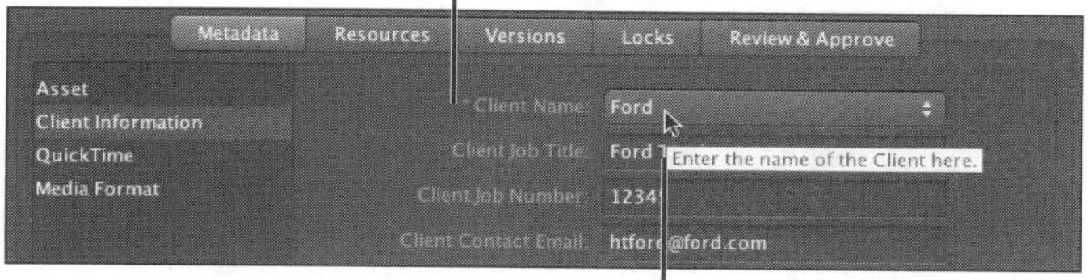

Notice that the field description shows up as a tool tip when the user's mouse rolls over the field.

Modifying Specialized Metadata Groups

Metadata groups are also used to describe which fields are used for tasks such as advanced searching, subscriptions, and even displaying search results.

To find specialized metadata groups, it's best to use the advanced search feature within the Metadata Group pane. For example, an advanced search for Category Equals Asset Metadata reveals metadata groups that help organize asset metadata throughout the program.

Perform an advanced search for specific categories to find corresponding metadata groups. In this case, all groups relating to assets are returned.

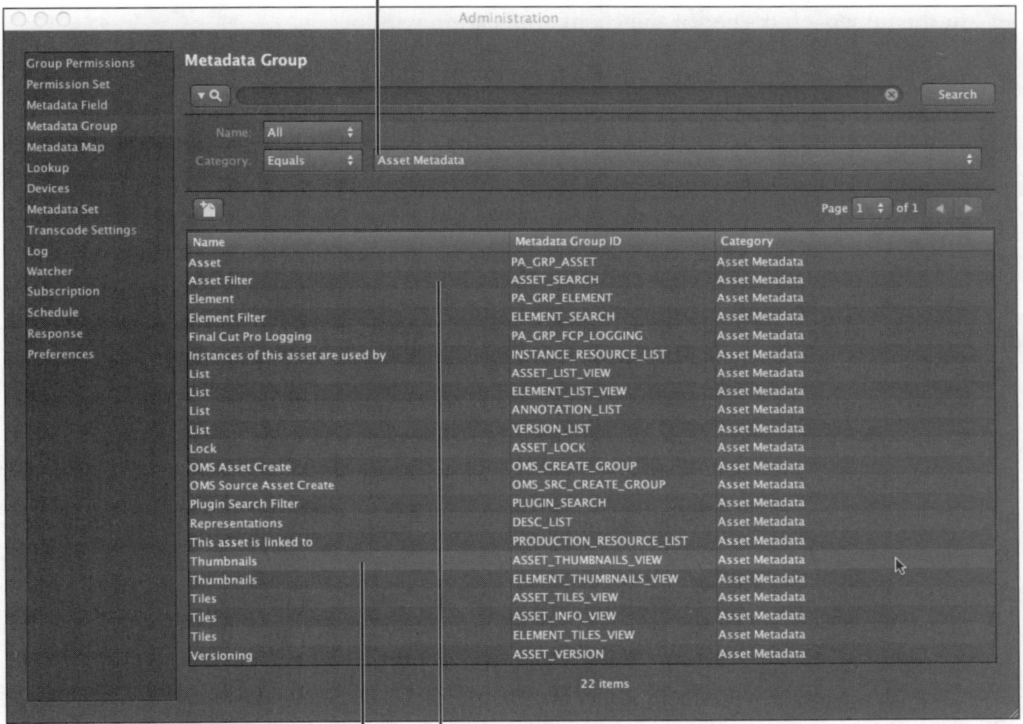

The first exercise of this section alters what fields are shown below the thumbnail view of a search result.

The second exercise modifies the Asset Filter to allow end users to search for assets based on Client Name.

Let's first alter the ASSET_THUMBNAILS_VIEW metadata group, which dictates what fields are shown when looking at the thumbnail view of a search result.

1. Double-click the ASSET_THUMBNAILS_VIEW listing.

2. In the Fields section of the window, add a few of your new metadata fields to the selected area at left.

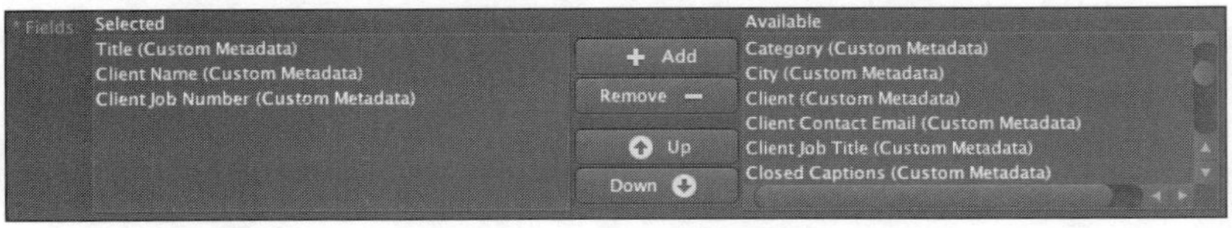

3. Click Save Changes at lower left.

4. Create a new workspace by selecting New Workspace from the Window pop-up menu in the upper-left corner of the client application window.

5. Perform a new search for assets, making sure you are viewing them in thumbnail view.

   You should now see the new metadata fields listed at the bottom of the thumbnails.

Is it clicking now? Altering view-based metadata groups allows you to customize the look and feel of the program, enabling your users to quickly see the information they need to work more efficiently.

This next exercise will alter the ASSET_SEARCH metadata group, one of the most widely used in the program, to change what assets are used for advanced searches. Later on, we will leverage the changes made to this group to create subscription-based automations and even modify permission sets for specific groups.

1. Back in the Metadata Group pane of the Administration Panel, use its advanced search to return the same list as we did the first time, using Category Equals Asset Metadata.

2. Find the ASSET_SEARCH group and double-click it to modify it.

3. In the Fields section of the window, add the Client Name metadata field to the selected area at left.

4. Once it has been added to the Selected list, click on the Client Name field and make sure to assign the Client Name lookups in the Field Properties area, as seen here.

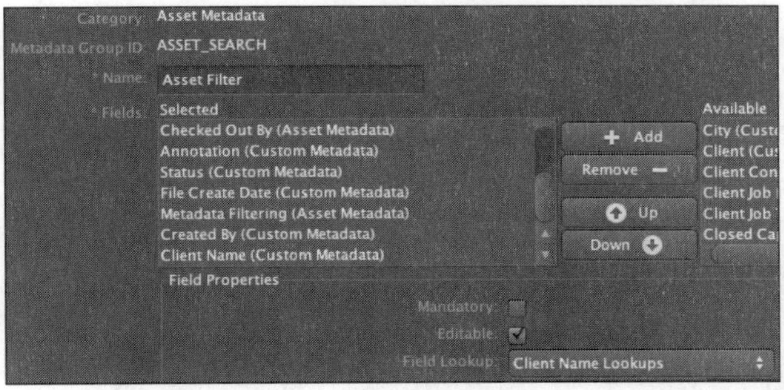

5 Click Save Changes at lower right.

6 Create a new workspace by selecting New Workspace from the Window pop-up menu in the upper-left corner of the client application window.

7 Perform a new advanced search in this window, and notice the new Client Name field shown at the bottom.

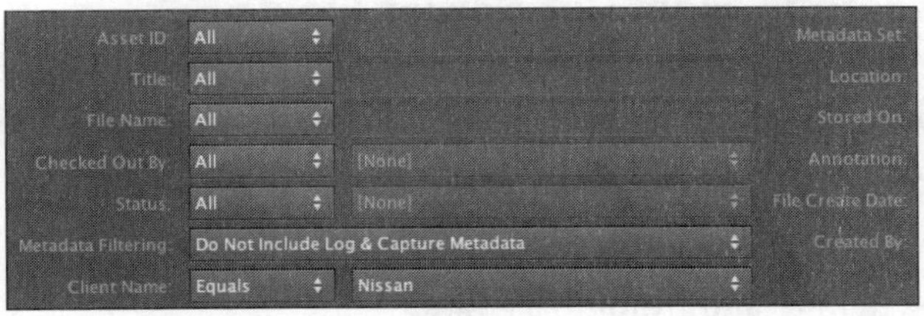

Now users can filter search results based on Client Name, which will greatly reduce the results returned, and further, help them narrow down the asset they need. Later in this lesson, we'll use this same change to create subscriptions and alter permission sets.

Creating and Editing Metadata Sets

The metadata sets available to you in Final Cut Server are based on the Customer Profile you selected when first installing the software. You may wish to expand these in order to categorize special assets and productions specific to your organization.

For example, the Video Production Customer Profile includes two metadata sets to describe Productions: Package and Show. Your organization might need a different set to describe the advertising spots that it regularly produces.

Use the Create Item button found in the Metadata Sets tab to create a new metadata set for an Advertising Spot.

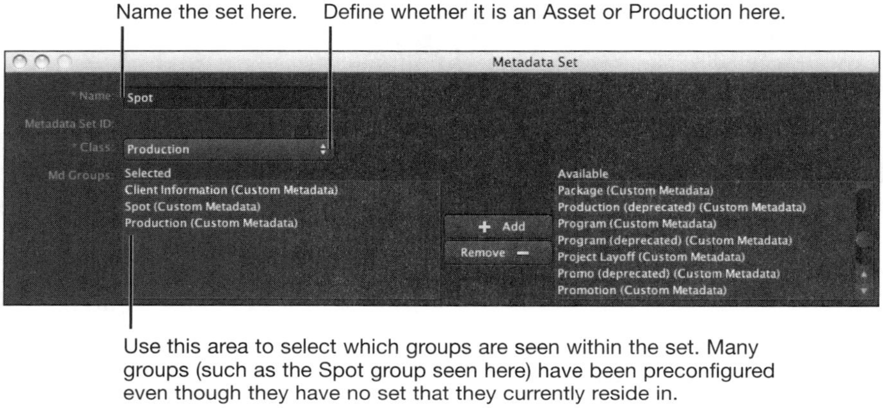

When saved, this new Spot set can be used immediately to create new productions.

Mapping External Metadata to Final Cut Server Fields

Metadata maps are used to specify which metadata fields from incoming files are placed into which corresponding fields in Final Cut Server. Similarly, these maps also define how Final Cut Server fields are placed into the metadata fields of outgoing files as well.

A good example of this is how Final Cut Server can map Adobe's XMP (Extensible Metadata Platform) metadata system to and from image files.

A quick search on keywords within the Metadata Maps pane will reveal a pair of one-way maps for XMP keywords into Final Cut Server's Keywords field, and vice-versa.

From Field	To Field
Keywords (Custom Metadata)	Keywords (XMP Metadata)
Keywords (XMP Metadata)	Keywords (Custom Metadata)

This allows any keywords within an image file (which can be added by selecting File > File Info in Adobe Photoshop) to be mapped into Final Cut Server.

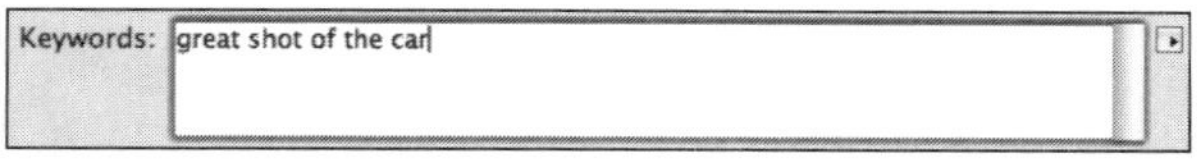

Creating keywords in Photoshop.

The same keywords mapped into Final Cut Server.

Many maps are already done for you. Besides XMP two-way mapping, there are also two-way maps for QuickTime file metadata and other file formats.

Sometimes, you may wish to create a map between a field from a file and a custom field that you have created in Final Cut Server. To do this, simply create a new two-way map, and make sure its priority is higher than any map that may want to use the same field—the lower the number, the greater the priority, with zero being highest.

In the example below, we are mapping the Creator Contact Info Email address from the XMP metadata in the image out to the Client Contact Email field we created in a previous exercise. Enabling the two-way map checkbox ensures that if we export an image from this asset, then the reverse map would occur.

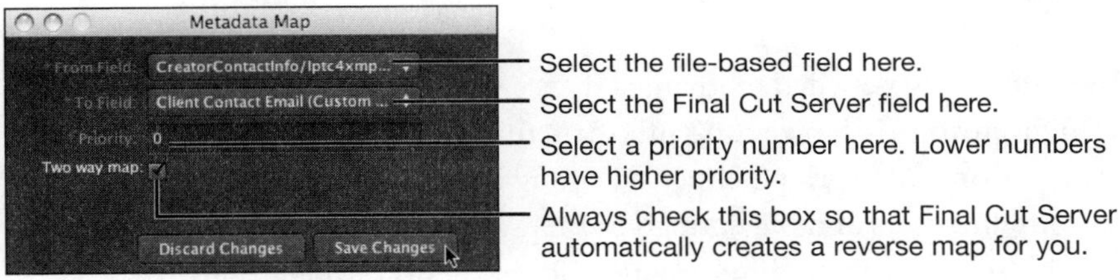
Metadata Map
From Field: CreatorContactInfo/Iptc4xmp...
To Field: Client Contact Email (Custom ...
Priority: 0
Two way map:
Discard Changes
Save Changes
Select the file-based field here.
Select the Final Cut Server field here.
Select a priority number here. Lower numbers have higher priority.
Always check this box so that Final Cut Server automatically creates a reverse map for you.

Giving Devices Edit-in-Place Status

It's best to always start defining a device with the Device Setup Assistant within the Preference pane, as explained in "Adding a New Device" in Lesson 7. You can then come to the Devices tab in the Administration Panel to further configure the device.

Devices
The Devices tab

Most of the additional settings that are found when editing Device settings are explained in detail within the Setup and Administration Guide that came with your software.

However, what is worth explaining here is the value of assigning edit-in-place status to a device, since the Device Setup Assistant in the Final Cut Server System Preference pane will not do this unless you are defining an Xsan volume.

A device that has edit-in-place status simply means that users can access that device's files directly rather than having to cache them locally before use.

This makes perfect sense with Xsan volumes, in the case where the Final Cut Server and the user's computer are both clients of a SAN, because it would waste time and hard drive space to download a file that the user can access directly off of the SAN. If a user is not a SAN client, then Final Cut Server falls back to locally caching files as an alternate way to give the user access.

But suppose the workflow of your facility requires users to modify content that is located on a powerful network file server. If regular manipulation of the file from its location on the file server is needed, then it might make sense to enable edit-in-place for that device.

Other Ethernet-based file systems, such as those from Isilon Systems, EditShare, or Omneon, will also need edit-in-place status because these systems are designed to offer files at very high data rates.

> **NOTE ►** Carefully evaluate whether giving a device edit-in-place status can be accomplished given your organization's infrastructure. The practicality of an edit-in-place device relies on a network pipeline fast enough to provide files at a speed that is tolerable to your end users and big enough to keep up with all the activity. Check the Appendix of this book for further information about these considerations.

To enable a device for edit-in-place, double-click on its listing within the Devices tab of the Administration Panel.

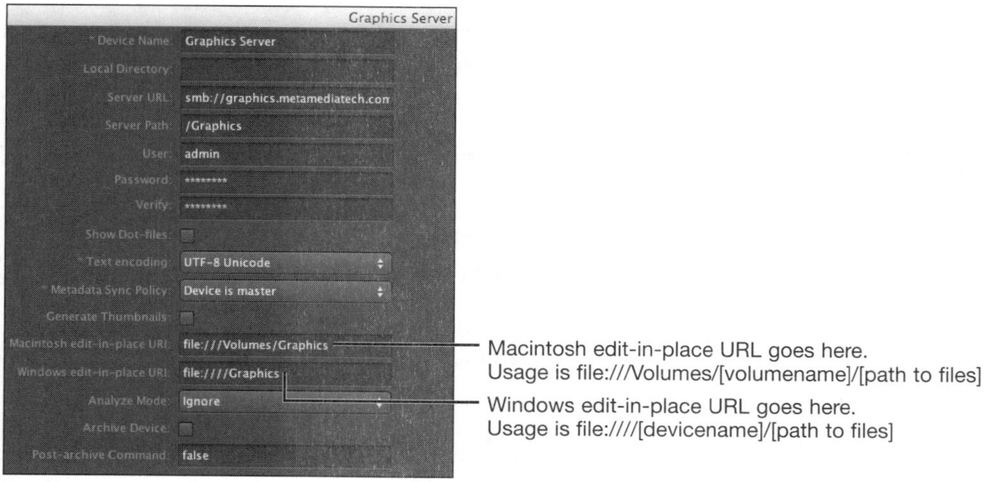

In the above example, after saving these changes, users who log in to Final Cut Server from a computer (Mac or Windows) with direct access to the Graphics file server should see all of its assets as immediately available.

Defining Transcode Settings

Transcode settings allow Final Cut Server to convert media during an upload, an export to an end user's desktop, or a copy response within an automation. Transcode settings are created and modified by clicking on the Transcode Settings tab, which brings up the Transcode Settings pane.

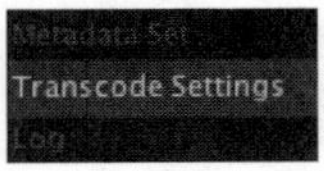

The Transcode Settings tab

Final Cut Server differentiates between two kinds of transcoding: Media transcoding for video and audio files and still image transcoding for image manipulation. We'll address each in a separate section.

Working with Media Transcode Settings

Apple's Compressor application comes with Final Cut Server, and it handles all media transcoding. All default transcode presets that come with Compressor are therefore available as transcode settings within Final Cut Server.

If your organization requires custom transcode settings, use the Compressor application located on your Final Cut Server to create a new Compressor setting, and then use the Transcode Settings tab in the Administration Panel to link a Final Cut Server transcode setting to the setting that you created in Compressor.

Creating Custom Settings in Compressor

Creating a custom setting is accomplished by either creating a new setting or duplicating an existing one within the Settings tab in Compressor.

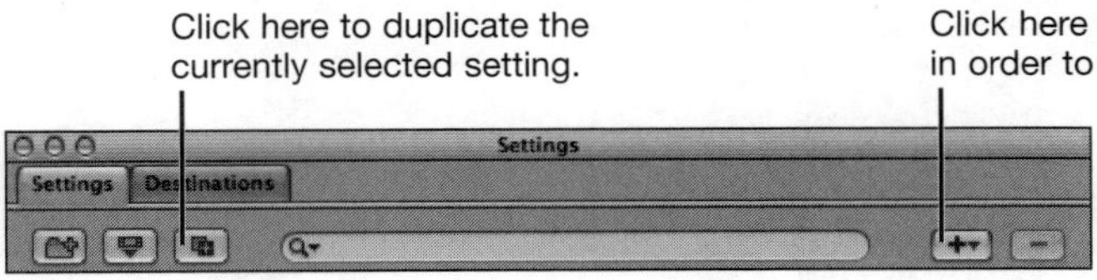

Click here to duplicate the currently selected setting.
Click here to pick from a list of formats in order to create a new setting.
Settings
Settings
Destinations

You then use the Inspector window to name and reconfigure the setting. Once saved, quit Compressor.

> **MORE INFO ▶** For more detailed information on how to create Compressor settings, refer to the documentation PDF files that come with your Final Cut Server software. Another fantastic resource for this is Brian Gary's *Apple Pro Training Series: Compressor 3 Quick-Reference Guide.*

Creating Settings for Other Industry Formats

Compressor utilizes a plug-in architecture that allows for expansion from third parties. If your organization needs to create transcodes in Final Cut Server using formats not covered natively by Compressor, then consider using a plug-in such as Telestream's Flip4Mac or Episode series products.

The addition of Flip4Mac Studio Edition provides instant Windows Media export for Compressor. Episode Pro's settings for transcoding into formats such as Flash and GXF are available as a plug-in for Compressor, which then links to Final Cut Server transcode settings, explained in the next section.

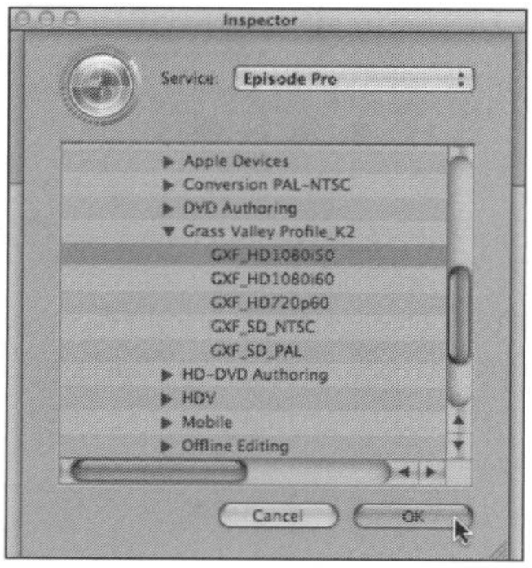

> **MORE INFO ▶** More information on Episode Pro can be found at http://flip4mac.com/episode.htm.

Linking Compressor Settings to Final Cut Server Transcode Settings

In the example below, we will create a new transcode setting for an MPEG-2 transport stream. This setting is a default setting within Compressor, so we can begin the task of linking this setting within Final Cut Server.

1 Click the Transcode Settings tab, at left, to open the Transcode Settings pane.

2 Click the Add Item button to reveal the Transcode Settings window.

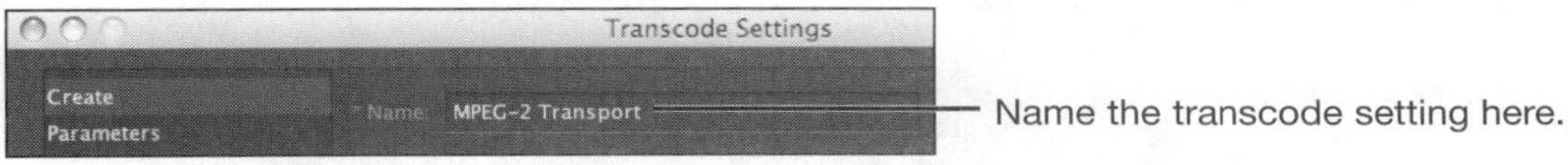
Transcode Settings
Create
Parameters
* Name: MPEG-2 Transport
Name the transcode setting here.

3 Click on the Parameters tab on the left of the Transcode Settings window to create the link.

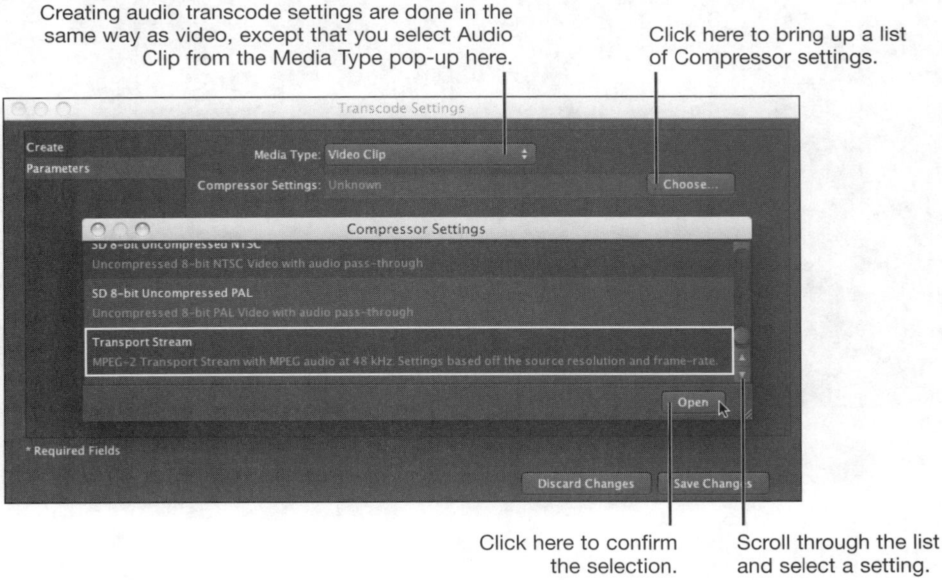
Creating audio transcode settings are done in the same way as video, except that you select Audio Clip from the Media Type pop-up here.
Click here to bring up a list of Compressor settings.
Transcode Settings
Create
Parameters
Media Type: Video Clip
Compressor Settings: Unknown
Choose...
Compressor Settings
Uncompressed 8-bit NTSC Video with audio pass-through
SD 8-bit Uncompressed PAL
Uncompressed 8-bit PAL Video with audio pass-through
Transport Stream
MPEG-2 Transport Stream with MPEG audio at 48 kHz. Settings based off the source resolution and frame-rate.
Open
* Required Fields
Discard Changes
Click here to confirm the selection.
Scroll through the list and select a setting.

4 Click Save Changes to save the transcode setting.

Creating Custom Still Image Transcoding Settings

Final Cut Server handles still image transcoding differently from video and audio, as it actually performs the process internally rather than handing it off to Compressor.

Most still image transcodes are done to extract frames from a video file, in order to create poster frame or thumbnail representations.

The best way to describe the creation of new still image transcode settings, therefore, is to take a look at one that exists. We'll take a look at the JPEG Thumbnail Proxy setting that creates the thumbnail images that you see within a search result.

Use this section to choose the file wrapper for the still image.

The Tracks section defines which tracks of the video file are used to create the still image. In this case, we are using the video track of a QuickTime movie file in order to derive the thumbnail.

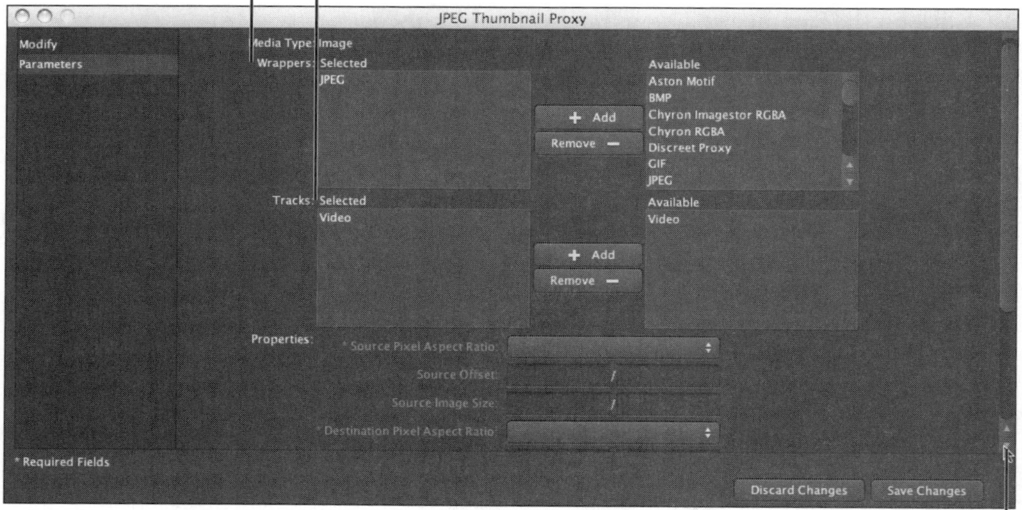

Scroll down to access further image manipulation settings.

After setting wrapper and track settings, scrolling down reveals further image manipulation settings.

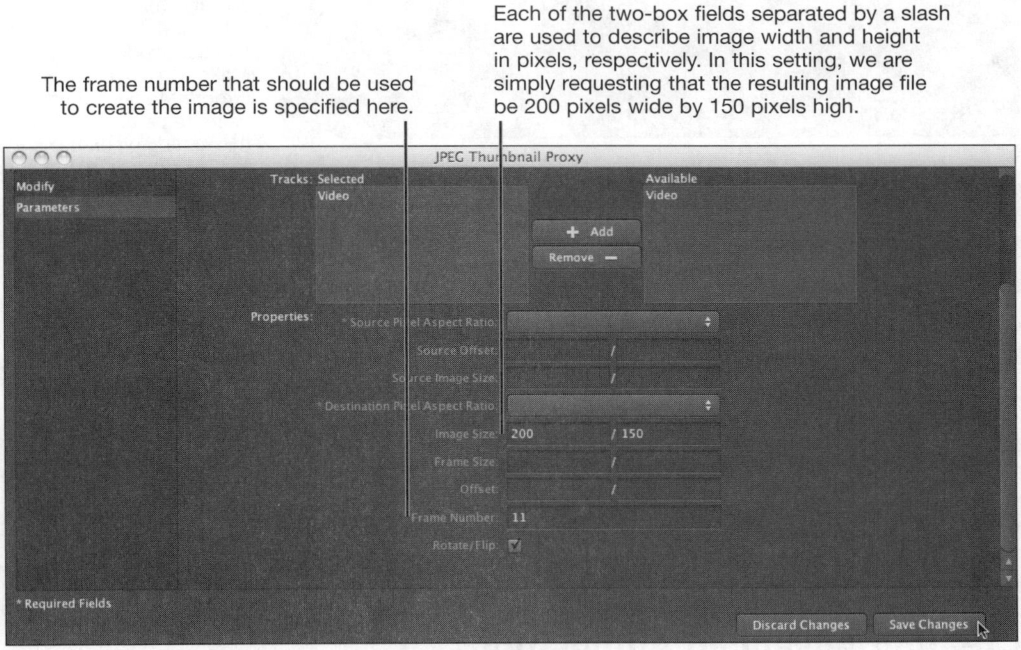
The frame number that should be used to create the image is specified here.
Each of the two-box fields separated by a slash are used to describe image width and height in pixels, respectively. In this setting, we are simply requesting that the resulting image file be 200 pixels wide by 150 pixels high.
JPEG Thumbnail Proxy
Modify
Parameters
Tracks: Selected
Video
Available
Video
Add
Remove
Properties:
* Source Pixel Aspect Ratio:
Source Offset:
Source Image Size:
* Destination Pixel Aspect Ratio:
Image Size: 200 / 150
Frame Size:
Offset:
Frame Number: 11
Rotate/Flip:
* Required Fields
Discard Changes
Save Changes

Creating a new still image transcode setting starts by creating and naming a new setting as in the previous section. Then, click on the Parameters tab, at left, and select Image from the Media Type pop-up menu. The window will then change to show settings for image manipulation, in the same manner as the pre-built setting we just explored.

Linking Settings to Devices

Once you have saved the new transcode setting, you *must* then search for it within the Transcode Settings pane and re-open it to specify which devices can use the setting when transferring files to or from it.

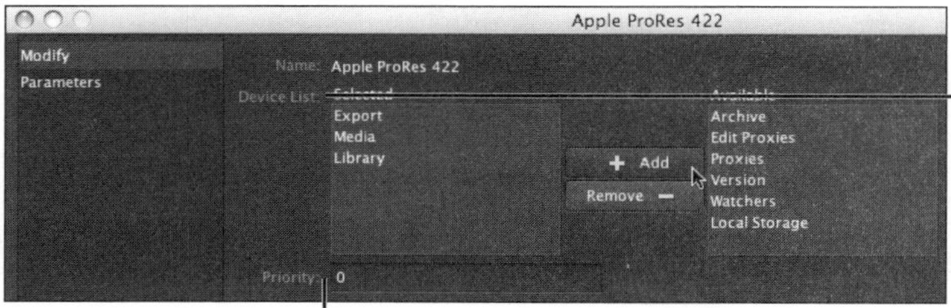

Use this section to specify which devices, listed at right, should use this transcode setting. Make sure to include Export if you want your users to be able to use this setting when using the Export feature in Final Cut Server.

Use the Priority setting to control how this setting will appear in a list of transcode settings for any of the above selected devices. Lower numbers give higher priority.

NOTE ► Notice in the image above that another device, called Export, is listed. This device does not show up in the Devices pane. The Export device is symbolic: it represents the action of a user choosing the Export command from an asset's shortcut menu. If you want to allow your end users more transcode choices in the export function, make sure to include the export device when linking transcode settings to devices.

Save your changes after your device selection is complete.

Creating Advanced Automations

The next sections explore the incredible power that is available with the Administration Panel when it comes to creating automations. All of these functions use the Watcher, Subscription, Schedule, and Response tabs within the Administration Panel.

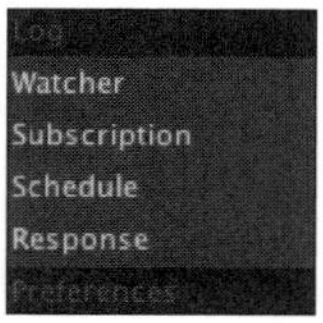

The tabs used to create and modify automations.

Automations are comprised of triggers and responses. There are three triggers in Final Cut Server:

- *Watchers*, which "watch" for new or modified media that appear inside of a folder contained on a device. A watcher is used most often to add and catalog media files. Most often, watchers are placed in locations that are accessible to the Internet, to other automation systems, or other "outer edges" of your organization's workflow.

- *Subscriptions*, which look for changes in the metadata of an asset, production, job, or system event. Subscriptions, combined with customized metadata, form the basis of the intelligent automations tailored for your organization. Examples of subscriptions include email notifications to the administrator when a job fails and the delivery of a transcoded file to a specific location based on the status of an asset.
- *Schedules*, which fire off at a certain absolute or periodic time frame. These events usually start scans that add and delete assets. They can also execute maintenance tasks such as clearing logs.

There are far more responses than you are exposed to in the System Preference pane. No less than 20 kinds of responses are available to attach to a trigger. The following table lists available responses and their typical triggers.

Response Definitions

Response	Function	Usually Triggered By
Check the disk space of the database volume	Creates a log entry if the remaining disk space on the database's hard drive is getting low.	Schedules
Clean Jobs	Archives and then deletes the Jobs list.	Schedules
Clean Logs	Archives and then deletes the log.	Schedules
Copy	Copies a file to a device, with or without transcoding. Optional provision to create an asset and specify its metadata set during the process.	Watcher or Asset Subscription
Delete	Deletes the item. The Always Run option allows a deletion even if there is a failure with one of the responses above it in a list.	Watcher or Asset Subscription
Email	Sends an email based on specified settings. You can place fields of an asset or production in the subject or body by inserting them in [brackets].	Any

table continues on next page

Response	Function	Usually Triggered By
Log	Reports this automation as an entry in the log.	Any
Measure database size	Determines the disk space used by the database. Reports this as an entry in the Log tab.	Schedules
Monitor Scan	Internal response. Not used in automations.	N/A
Move to Archive	Moves an Asset's primary representation to the selected archive device.	Asset Subscription
Purge Subscriptions	Internal response. Not used in automations.	N/A
Read XML	Reads data from an XML file discovered during a watch operation. (See Write XML.) Takes information from that file to perform a Set Metadata response (also in this table) for a current asset.	Watcher
Restore from Archive	Restores the primary representation of an asset from the archive device on which it was originally placed. Deletes the archived file once the primary representation is back where it was.	Asset Subscription
Run an external script or command	Executes a shell script or command-line binary. You must specify the path for the command. You can also specify arguments for the command.	Any

table continues on next page

Response	Function	Usually Triggered By
Scan	Scans a specified device. Usually used when the device's files have never been cataloged or are being changed outside of Final Cut Server. Full scans both add and delete assets based on the results of the scan. Add only scans just add new assets based on new files that are discovered.	Schedules
Scan Productions	A scan that can also place discovered file's assets into a new or existing production, and can also place path information into metadata fields within the production.	Schedules
Search Expired	Searches for assets that meet certain time-based criteria. Used with custom event definitions (Custom 1, 2, or 3) to then create a subscription of that event to trigger further archive and/or delete responses.	Schedules
Set Asset Metadata	Changes the metadata fields within a single metadata group of an asset.	Asset or Production Subscription
Set Production Metadata	Changes the metadata fields within a single metadata group of a production.	Asset or Production Subscription
Write XML	Outputs all of an asset's metadata within an XML file and places it in a specific path. Used in conjunction with the Read XML response.	Asset Subscription

When creating an automation, it's helpful to first plan out what needs to happen, and then plan what needs to be true in order for that to happen. Once you have a clear idea of what you want to do, you'll actually be creating the automation *backwards*. Unlike creating automations in the System Preference pane, in the Administration Panel, you'll be defining responses *first*, and then attaching them to a trigger *second*.

It's comforting to know that watchers and subscriptions first created in the System Preference pane can be further modified within the Administration Panel. So you might want to create those kinds of automations first within the System Preference pane, and then come to the Administration Panel to have access to more settings.

Defining Responses and Response Details

This next section adds clarity to the responses that aren't immediately intuitive. Of course, more detail on all responses is available in the Setup and Administration Guide that came with your software.

To create a new response within the Administration Panel, click on the Response tab to reveal the Response pane. Then click the Create New Item button. In the Name and Description fields, name and describe the function of the response. The more detailed a name and description you give, the easier it will be for you and other administrators to understand your intentions when accessing it later on.

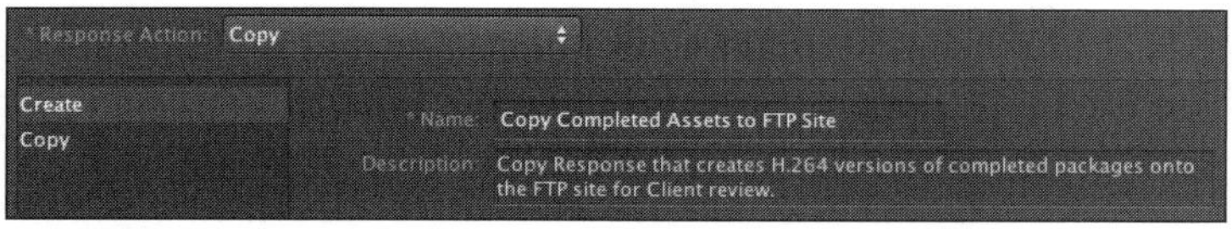

Copy

The Administration Panel provides a few more options for the copy response than are given to you within the System Preference pane.

Select whether this copy response has higher priority than other jobs that are currently being processed. The default value is 2.

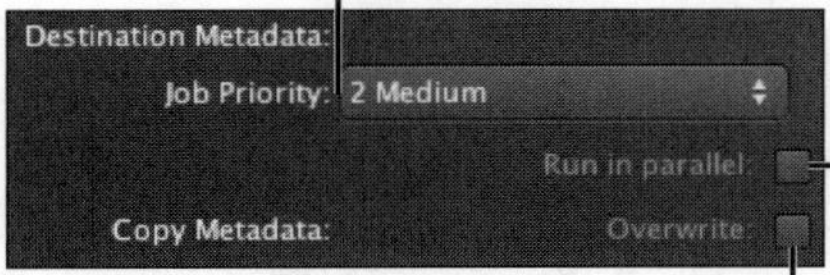

Click here to have the copy response run at the same time as other jobs. This is not recommended unless you have a very powerful computer running Final Cut Server and your storage can handle simultaneous I/O operations.

Click here to take all metadata from the subscribed or watched asset and place it within the new asset created from this copy response.

When setting up a copy response here, you have the choice of whether to create a new asset from the copied media. You can therefore also elect to associate that asset with a production at the bottom of the window.

Associate with Production option.

Read and Write XML

These Read and Write XML responses work in conjunction to provide metadata access and modification opportunities outside of Final Cut Server.

An example workflow for using these two responses would be as follows:

1. An asset or production subscription would trigger a Write XML response to write out that asset or production's metadata into an XML file to a specific path.
2. An external program or script would scoop up that XML file, parse its information, and present it within an application, web page, and so on.
3. A user would be able to view and perhaps change this metadata.
4. The external program or script would then generate a custom XML file, as in the example that follows, with the changes to the metadata, and place it in a specific watch folder.
5. Final Cut Server would be watching that folder for XML files, trigger a Read XML response, and set the original asset or production's metadata with the information contained within the XML file.

The Write XML response simply outputs an XML encoded file that contains all of an asset's metadata to a particular device location. All you do in this response is specify a path for the file.

Write XML is *always* used as a response to an asset subscription. Because of this, the output file's name will be the same as the title of the asset, with an .xml extension. (If the asset's Title is kitten, the resulting file from this response will be kitten.xml.)

You can then pick up that metadata using a program or script that you create outside of Final Cut Server. Future third-party applications will also take advantage of these XML files to display Final Cut Server asset metadata in their applications.

The Read XML response is *always* attached to a watcher that is looking specifically for XML files. There are no settings for the Read XML response, since all it will do is try to read XML files that are placed in a watcher that you set up. The XML file you supply must be in a specific format, containing information to perform the equivalent of a Set Metadata response, which modifies or inserts information into fields of a metadata group. A sample of the format of the file is provided here.

```
<?xml version="1.0"?>
<FinalCutServer>
<request reqId="setMd" entityId="/asset/20">
<params>
<mdValue fieldName="Status" dataType="string">Approved</mdValue>
<mdValue fieldName="Keywords" dataType="string">flame fire</mdValue>
<mdValue fieldName="Web Comments" dataType="string">this is what we were looking for</
mdValue>
</params>
</request>
</FinalCutServer>
```

Of particular importance to the file above is the *entityId*, which is the asset or production identification number, originally assigned to the asset when it was made. This must be present and accurate for the Read XML response to function properly.

TIP The file placed into a watcher that has a Read XML response must be formatted precisely as shown above. This means that any metadata field that you are populating or replacing has to be spelled exactly as it currently exists within Final Cut Server's catalog. To get a good idea of how to create the Read XML file for the round trip, create a simple subscription to a few assets that generate a Write XML response, and then take a look at the .xml files that come out. This will give you a clear idea about which metadata fields can be manipulated outside the program and how to write the file that will be used in the corresponding Read XML response.

Scan

You've probably encountered a Scan when first setting up a device within the System Preference pane, where both the frequency and type was specified in one step.

In the Administration Panel, a Scan is a response generated from a Scheduled Event, and as such, you can have much greater control over the kind of Scan and when it occurs.

The first few controls are similar to what's found in the Preference Pane with a few more choices.

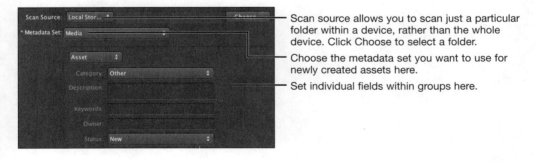

Scan source allows you to scan just a particular folder within a device, rather than the whole device. Click Choose to select a folder.

Choose the metadata set you want to use for newly created assets here.

Set individual fields within groups here.

When you scroll down, the most powerful option here is to filter for both what you want to scan and what you don't want to as well.

Select either Add Only (just add new files to the catalog), Full (add new files and also delete assets for which files no longer exist), or Purge Only (just delete assets for files that no longer exist) from this pop-up menu.

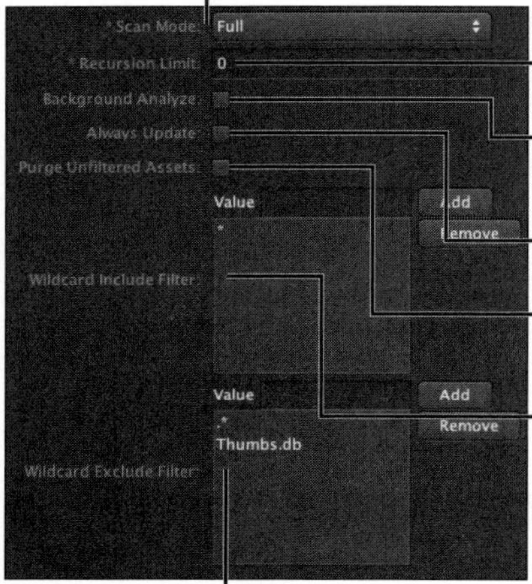

Recursion Limit sets how many folders "inward" to scan from the starting location. A value of 0 means unlimited.

Click here to process discovered files in the background. Not recommended unless you are running Final Cut Server on a very powerful Mac.

Click here to reanalyze assets even if they have not changed since the last scan.

Click here to purge assets for which original representations have been deleted, even if they don't meet the criteria of the filters below. Only needed for a full or purge-only scan.

The Wildcard Include Filter will include only files who meet the name-based criteria you enter in this box. You can place filename extensions, as well as any partial filename that should be included. Remember to include the asterisk if you want to include all files.

The Wildcard Exclude Filter works exactly opposite from the Include Filter. In this example, we're excluding files that are often found on Mac (HFS+) filesystems but should be ignored, such as hidden files (ones that begin with a period) and the thumbnail database file located in every folder in OS X.

Scan Production

Scan Production provides all that a regular Scan does, with the addition of being able to add all assets into a specific production. The additional prize here is that you can take path information and insert it into metadata fields of the production. This is done by placing the folder's hierarchical number within brackets into the field, as is shown here.

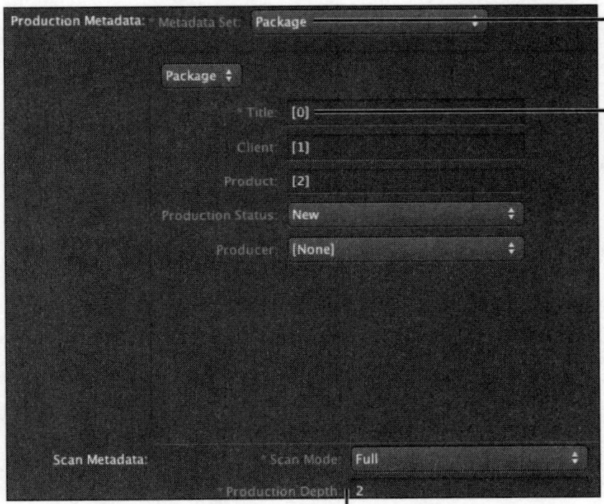

Set the metadata set for the production from this pop-up menu.

Numbers within brackets assign folder names from the file path to specific fields. In this example, the root folder's name gets placed in the Title field, the folder inside of the root folder's name gets placed in the Client field, and so on.

Production Depth determines the number of folders that are searched "inward" from the root folder of the Scan in order to provide metadata within production fields, as explained above.

Run an External Script or Command

This response is one of the most powerful in Final Cut Server, because it allows the script-savvy administrator unlimited flexibility to do further actions outside of the program.

If this response fires off due to a subscription to an asset or production, arguments can pass to the script (or binary command), which contain metadata fields from that subscribed asset or production. In similar fashion to an email response, these fields are placed in brackets, which allows the external script or command to use the information within the fields.

However, you can simply create schedules that fire off external commands at a certain time of day (or night).

For example, you could back up your proxies device on a nightly basis to a mirror volume by firing off an external rsync command. If the full command text is

```
/usr/bin/rsync –av /Volumes/PROXIES/Proxies.bundle /Volumes/PROXIESMIRROR/
```

then you would insert the actual binary command or script in the Command Path field, and all of the arguments in the Command Parameters field, as shown here.

After creating this response, you then attach this response to a schedule (explained below) to make sure that your proxies device is synchronized to its mirror every night.

Search Expired

This response allows you to automatically archive and delete assets that have reached a certain "age." It works in two parts.

First, the Search Expired response is triggered by a schedule. When it fires, it combs through the catalog, looking for assets that meet its criteria. When they do, one of the three Custom events in Final Cut Server (Custom 1, Custom 2, or Custom 3) is attributed to those assets.

Second, a subscription to that custom event (and perhaps a few more criteria, like assets which have primary representations on a particular device) triggers the responses you want to achieve on these assets, such as archive or delete.

In the next example, we have created a Search Expired response that is looking for assets that have not been accessed for the last 30 days. If this is true, those assets will have a Custom 1 event attributed to them. This Custom 1 event can then be subscribed to in order to trigger further responses such as email and archive.

Select one of the Custom event types here. Any time this response finds an appropriate asset, this event will be tagged so that a further subscription on this event will allow this asset to be singled out.

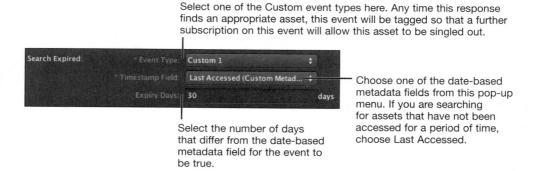

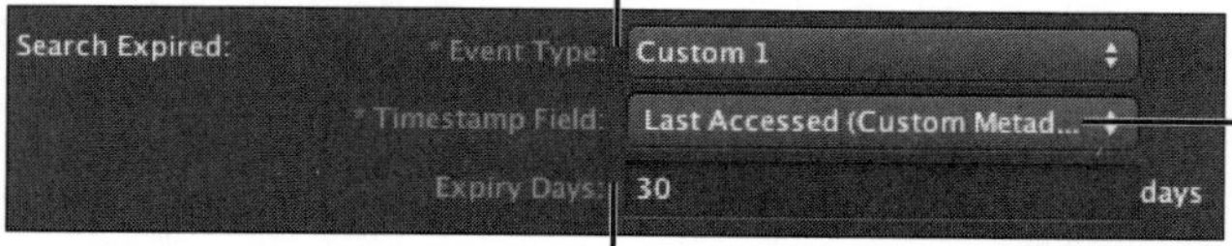

Choose one of the date-based metadata fields from this pop-up menu. If you are searching for assets that have not been accessed for a period of time, choose Last Accessed.

Select the number of days that differ from the date-based metadata field for the event to be true.

Defining Triggers and Trigger Details

Now that your responses are set, the final step is to configure the trigger for them. In all cases, triggers have some universal settings in the Create tab that are explained here.

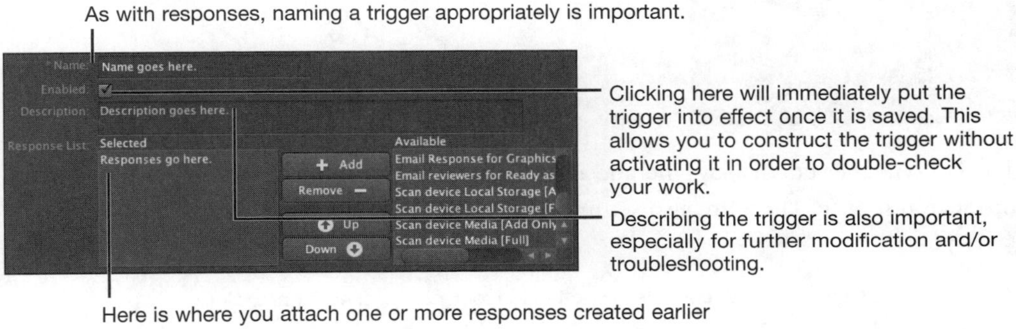
As with responses, naming a trigger appropriately is important.
* Name: Name goes here.
Enabled:
Description: Description goes here.
Response List: Selected
Responses go here.
Available
+ Add
Remove —
Up
Down
Email Response for Graphics
Email reviewers for Ready as
Scan device Local Storage [A
Scan device Local Storage [F
Scan device Media [Add Only
Scan device Media [Full]
Clicking here will immediately put the trigger into effect once it is saved. This allows you to construct the trigger without activating it in order to double-check your work.
Describing the trigger is also important, especially for further modification and/or troubleshooting.
Here is where you attach one or more responses created earlier

Schedule

When creating a scheduled event, you need to determine when it will occur. Select one of the four choices below, then click the Schedule tab at left to further specify the time. The Periodically setting is used to specify a trigger that fires off every number of minutes.

* Schedule:
Create
Daily
Hourly
Periodically
Weekly

Watcher

Don't be confused about the two kinds of watchers (poll and subscription), because poll is the only one you can use for automations. Poll watchers continuously look at the contents of a folder, measuring a file's size until a consistent number is returned for a certain number of cycles. The file is then used for the responses that are attached to the watcher. The Create tab has a few additional settings.

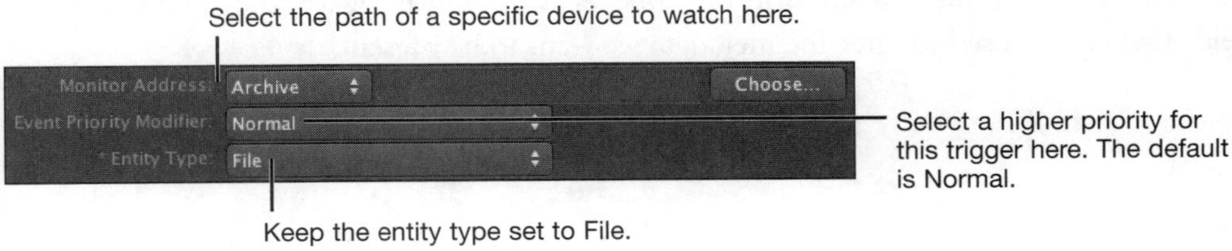
Select the path of a specific device to watch here.
Monitor Address: Archive
Choose...
Event Priority Modifier: Normal
* Entity Type: File
Select a higher priority for this trigger here. The default is Normal.
Keep the entity type set to File.

The Event Type Filter, located at the bottom of the Create tab, allows you to specify what kind of files should trigger the responses. Because this is a watcher, you should *always* select only Created, so that the watcher only looks for new files that wind up in the folder.

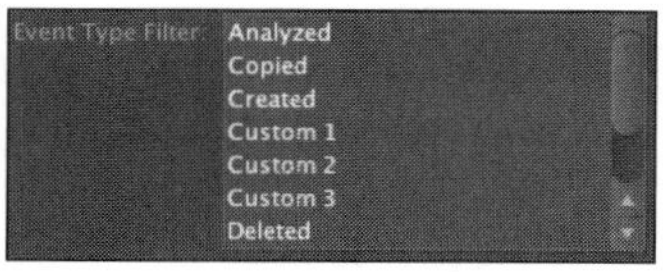

The Poll Watcher tab displays include and exclude filters identical to those explained in the Scan response. There are also settings to define how often the polling of the folder takes place.

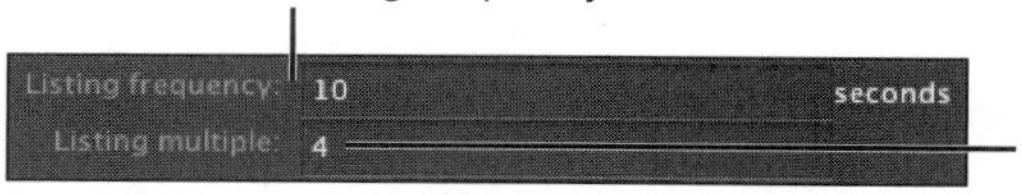

Listing frequency controls how often the folder is checked for new files.
Listing frequency: 10 seconds
Listing multiple: 4
Listing multiple controls how many times a file's size must return the same value before the file is considered complete.

Subscription

Subscriptions have essentially the same settings in the Create tab as the other triggers mentioned above. The Event Type Filter within a subscription can take on greater significance since we could use it to simply subscribe to a Custom event that we had triggered with a Search Expired response. Once the settings within the Create tab are set, we then select the kind of subscription we want to perform. The three choices are Asset, Job, and Production subscriptions. (System is not used for automations.)

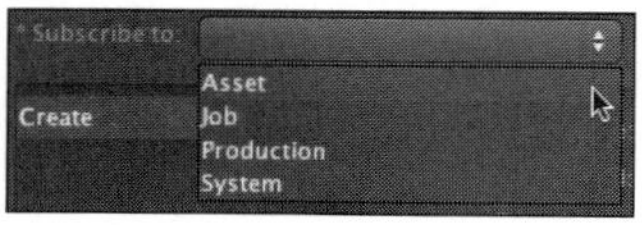

Once selected, a tab for that subscription appears at left. Once clicked, the tab displays fields that can be used to filter the metadata we want to be present.

For example, if you select an asset subscription, then whatever changes to the ASSET_FILTER metadata group were made (as we did earlier) will reflect in the fields you can use to create the subscription. Since the Client Name field is now part of this group, we can now subscribe to that field, as seen here.

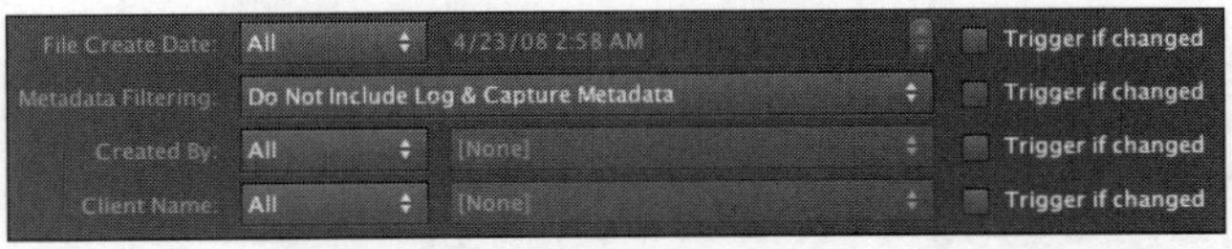

Of particular note is the "Trigger if changed" checkbox, which should be selected for at least one of the fields that you use in the filter. This ensures that the subscription will only fire off if the metadata for that field is *changed* to this state. Otherwise, items that had previously had a field with the criteria in question would fire off automatically.

Adding and Editing Permission Sets

Tools within the Administration Panel for managing groups and their permissions are contained within the first two tabs to the left. The Group Permissions tab, which is usually the first tab you'll see highlighted when you enter the Administration Panel, has the exact same function as its counterpart within the System Preference pane. For a refresher of how to use it, see "Defining Groups and Permission Sets" in the previous lesson.

The two tabs that configure Groups and Permissions Sets

The Permission Set tab is unique to the Administration Panel. Here, you can define new and modify existing permission sets. As you do, you can change the privileges of these sets, in terms of access to assets, productions, and functionality within the application.

Remember that when changing these settings, it's best to do so when no one is logged into the Final Cut Server using the client application. Off hours are therefore the most appropriate times for these changes.

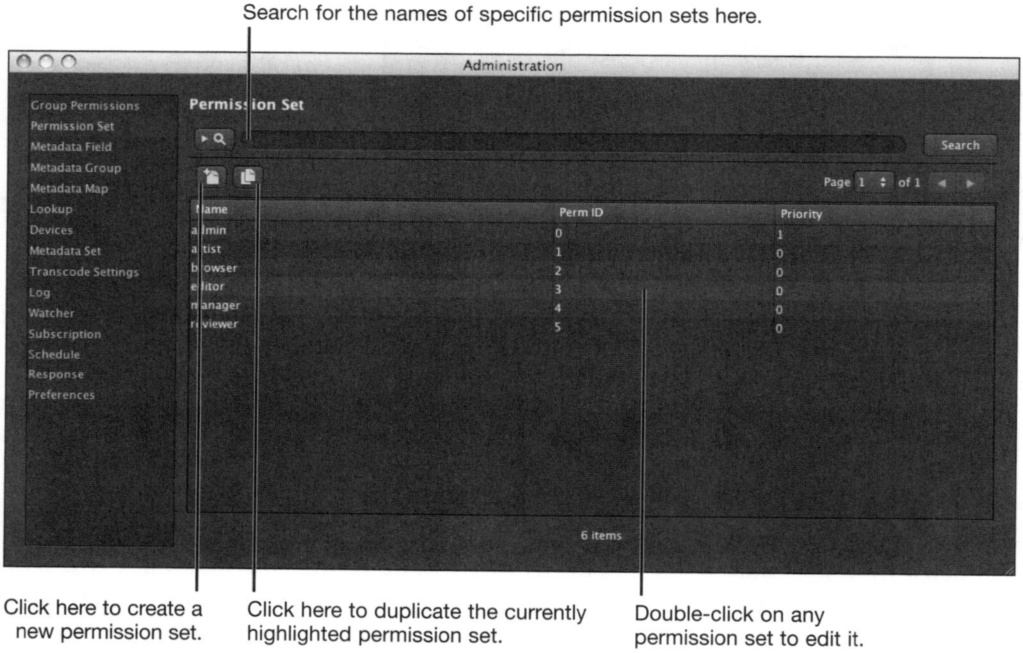

The Create, Duplicate, and Edit Permission Set functions are also available from the shortcut menu for each permission set.

Double-click a permission set (or select Get Info from its shortcut menu) to reveal the permission set detail view.

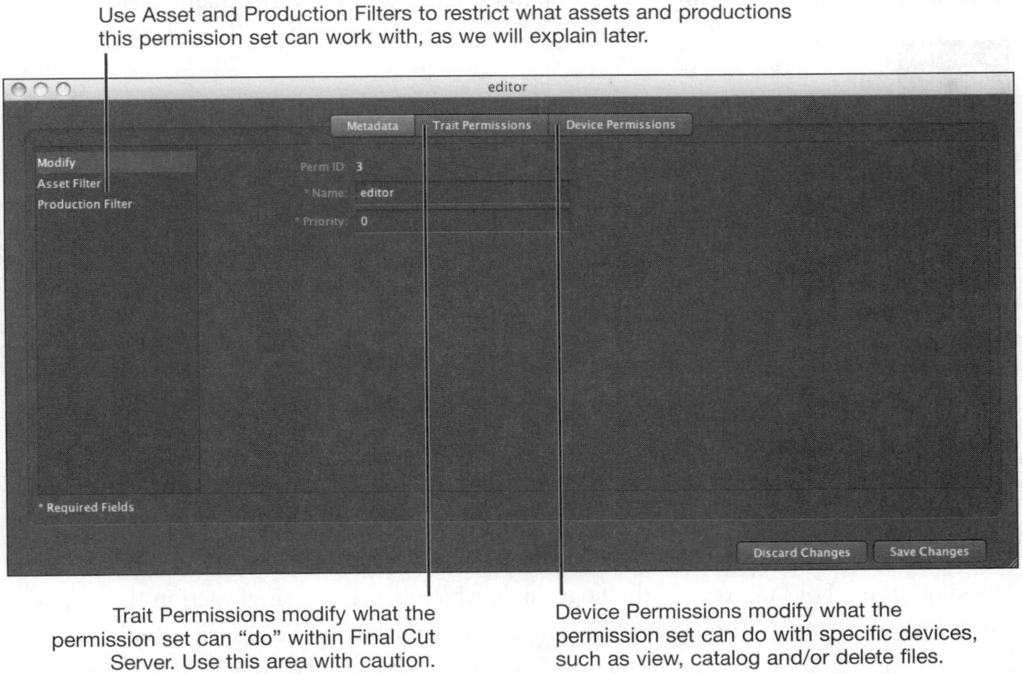
Use Asset and Production Filters to restrict what assets and productions this permission set can work with, as we will explain later.
editor
Metadata
Trait Permissions
Device Permissions
Modify
Asset Filter
Production Filter
Perm ID: 3
* Name: editor
* Priority: 0
* Required Fields
Discard Changes
Save Changes
Trait Permissions modify what the permission set can "do" within Final Cut Server. Use this area with caution.
Device Permissions modify what the permission set can do with specific devices, such as view, catalog and/or delete files.

The most likely task in the permission set detail view is to limit the kinds of assets and productions that can be seen by a particular permission set. Click the Metadata tab on the top and use the Asset and Production Filters at left to select available items.

Once again, since we learned previously to modify the Asset Filter (using the ASSET_FILTER metadata group), we have our Client Name field inside this filter. This will allow us to define a permission set that can only access assets that belong to one of our clients. Now we have an easy way to create permission sets for each client, so that when any member of that client's group logs in, they only see assets that apply to them (and not any other clients!).

If making changes on the Trait Permissions or Device Permissions tabs, each box in the matrix has three options: Inherit, Permit, and Forbid. The Inherit selection tells Final Cut Server to take the setting from the Permission Set that has a higher group priority than this one.

NOTE ▸ Review all of the current permission sets to see the differences between them. Most of the changes you'll see are within the Asset Filter and Permission Filter tabs. Then, begin changing these settings slowly, testing as you go, in order to modify the experience for a particular permission set.

Using the Logs Tab

The Logs tab

Clicking on the Logs tab reveals the Logs pane, which lists every single action that Final Cut Server performs (not just jobs that can be seen by viewing the Search All Jobs option in the Client Application window). Like all other tabs, it can be searched, and double-clicking on an entry brings up its detail view. A convenient feature of the Logs pane, as opposed to the Search All Jobs window, is that error details for failed jobs are immediately accessible in an info window, as seen here.

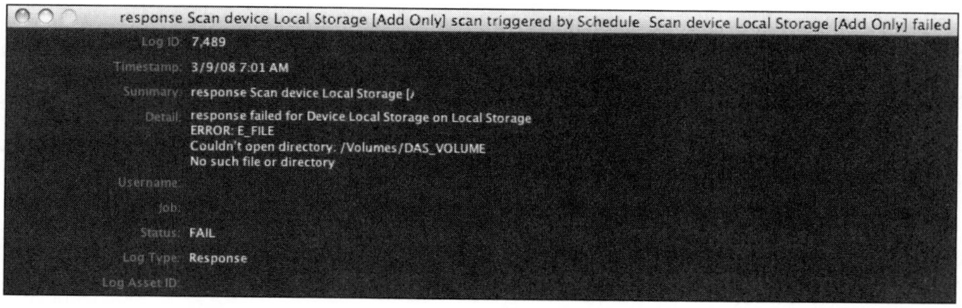

Using the Preferences Tab

The Preferences tab

The Preferences tab allows you to change system-wide settings. Each of the five sub-tabs of the Preference pane is explained here.

Global Preferences

The Global Preferences sub-tab mimics settings found in the System Preference pane (see Lesson 7), with the exception of one: Default Media Metadata Set, which determines what metadata set is used by default when uploading or cataloging new assets.

Compressor

Final Cut Server can be configured to use a Compressor cluster instead of the single instance of Compressor running on Final Cut Server. This sub-tab allows you to discover and connect to such a cluster, and enter the password for it if one is needed.

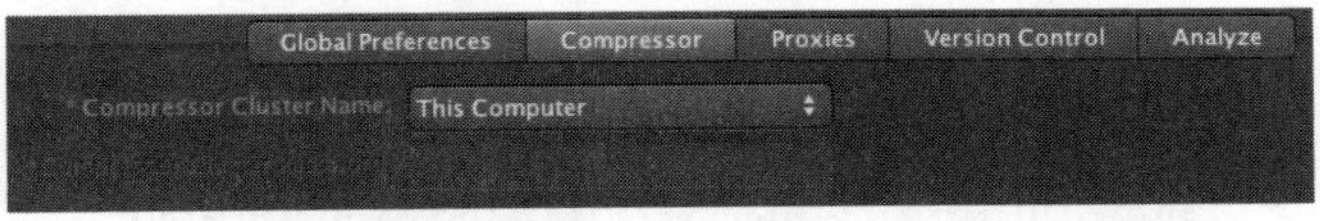

"This Computer" refers to the computer running Final Cut Server
(not necessarily the computer that you're running the client application on).

> **MORE INFO ▶** If you're interested in creating a true multi-machine cluster, or if you'd like to set up your single computer running Final Cut Server to be able to simultaneously process more than one Compressor operation at a time, see "About Creating a Custom QuickCluster" on page 33 of the Setup and Administration Guide.

Proxies

The Proxies sub-tab allows you to reconfigure the contentbases (see the following More Info for a definition) used for both standard and edit proxies. Unless you have a specific reason to do so, these should be left as is in order for you to access the proxies that have already been created. For example, changing these settings to another device will probably make all of the clip proxies in your catalog inaccessible!

MORE INFO ▶ Contentbases are devices, except they are used exclusively by Final Cut Server to store files for specific tasks like proxies and version control. They are formed as bundles to protect their contents from casual interaction in the Finder. For more information about contentbases, see the Setup and Administration Guide that came with your software.

The Enable Edit Proxies checkbox allows you to start creating edit proxies for your assets. As edit proxies use ProRes 422 as their codec, this feature would have benefit only for facilities that store uncompressed SD or HD video on their devices.

NOTE ▶ Edit proxies, if enabled, only get created when media files upload as part of a Final Cut Pro project being uploaded or checked in. Therefore, assets that become part of the catalog from scanning or a watch folder will not have edit proxies made for them.

Version Control

When version control is enabled on your Final Cut Server, the settings in this sub-tab configure the contentbase for the version control system. As with the Proxies sub-tab, the Control Device setting should be left alone. The Control Limit setting allows you to limit how many versions of an asset are saved within the version contentbase. The default value is 10.

Analyze

The Analyze sub-tab allows you to change the default transcode settings for both image-based and video-based representations. For example, suppose users don't have a lot of hard drive space, but they need to cache a lot of assets. They would therefore benefit from a lower transcode setting than the luxurious default H.264 for clip proxies. In this case, you can select a more efficient transcode setting for clip proxies, and all future clip proxies will be created at this setting. Remember that you if you want to use a new transcode setting here, it must first be created in the Transcode Settings pane, mentioned earlier in this lesson.

Lesson Review

1. Can you create custom video or audio transcode settings within Final Cut Server?
2. Where can you customize the advanced search area of the client application window?
3. Where in the Administration Panel would you change the default transcode settings for the various proxy formats when an asset is added to the catalog?
4. What is the relationship between a metadata field, metadata set, and metadata group?
5. What is a metadata lookup?

Answers

1. No, you must create them within Apple's Compressor application and then add them into Final Cut Server in the Transcode Settings pane of the Administration Panel.
2. In the Metadata Group pane of the Administration Panel, search for and edit the Asset Filter metadata group. This affects the fields that appear in the Asset Filter pane.
3. In the Analyze tab, which is in the Preferences pane of Administration Panel.
4. Every piece of metadata is stored in a metadata field. These are collected into metadata groups. Several metadata groups form a metadata set.
5. Lookups are pop-up menus that restrict the user's metadata choices. For instance, for the metadata field Status, specific stages of the approval process are listed so that the entire organization understands at which stage the process is.

9

Goals

- Search metadata tags based on logging information from Final Cut Pro
- Link various assets together in a production
- Use a watch folder to automate an approval process
- Build and save a Smart Search
- Revert to a previous asset version

Lesson **9**

Workflows and Final Cut Server

Now that you have seen Final Cut Server's capabilities, it's time to watch how Final Cut Server fits in with the daily workflow in your organization. Most production workflows, including your organization's workflow, share typical steps or processes. We will be looking at some of these shared processes in this lesson.

Because it would be difficult to demonstrate every possible configuration of Final Cut Server, we'll focus on two sample workflows that are likely to have features you can apply to your environment: a broadcast news television station and a medium-sized rich-media publishing company.

In the first example, a television station is changing to a digital file-based workflow. We'll explore how Final Cut Server facilitates the transition. The second example illustrates how Final Cut Server integrates with the existing workflow at a content-creation house to improve and enhance it.

Before we get to the examples, let's review the general steps for creating a workflow using Final Cut Server.

Planning Your Workflow

Final Cut Server will improve your organization's overall production process whether you're creating a new workflow or incorporating Final Cut Server into what you already have in place. You won't have to completely uproot your existing workflow or storage structures. However, a successful implementation will require careful planning to analyze the processes of your organization and make possible adjustments based on the advantages of Final Cut Server's tools. Use the following steps to help you design or improve your workflow:

- Analyze your workflow—Meet with key personnel from all departments that will interact with Final Cut Server as part of their workflow. Determine what specific processes they use. Use a simple flowchart or other diagramming tools to map out a development plan for your workflow, beginning with the immediate needs of your team's production process, keeping in mind possible future improvements. Especially important is the identification of redundant or cumbersome steps that Final Cut Server could help improve. It's not about corner cutting, but about recognizing parts of the process that are slowing your people down.
- Design your workflow—If you can, have a detailed discussion with a qualified Final Cut Server integrator about the basic needs and processes of your workflow. Determine how specific Final Cut Server tools can be used in your workflow and how to best configure them.

  Start simple by asking the following questions:

  Who are your users and what groups should they be in?

  Who needs access to which assets?

  Which storage units will be devices and what purpose will they serve (for example, media or archive)?

  Which automation capabilities will immediately help your workflow?

  What type of metadata information is important to your team?

  How will the metadata tags be grouped and structured?

  Metadata is a very powerful tool in Final Cut Server. Significant time should be spent early in the process to define the custom metadata fields, groups, and sets that your organization needs and how they will fit into your existing workflow.

  Add more complex features and capabilities as the demands of your workflow and staff require them.

- Train your staff—Proper instruction on the improved workflow process will enable staff to use Final Cut Server immediately. Comprehensive diagrams and documentation go a long way here. Make sure to focus training on the specific processes of each group in your organization, and how they fit into the entire workflow. People will be more motivated to participate in their processes when they see the importance of their contribution to the overall plan. Also, anyone acting as a Final Cut Server administrator will need specific training if he or she plans to make future changes to the workflow.
- Plan future workflow changes—After Final Cut Server is up and running, make adjustments only with careful consideration of their impact on the workflow. Make sure that your internal technical staff has detailed documentation of the overall Final Cut Server installation and configuration so that they can analyze how the effects of a change will ripple through the system. Begin by verifying the current processes in your workflow and determine what new features you wish to add to the workflow. Schedule a specific time when these changes can be applied without disrupting the production process in your organization. Then test your changes. The key here is to phase in the adjustments; you can seriously derail a process by making last-minute changes.
- Make it easy to communicate with each other—Make sure that any issues or suggestions from your team have a place to go—for example, an internal forums page on your company website. Regularly scheduled meetings, ideally scheduled prior to any planned updates to the Final Cut Server workflow, can also help people to effectively communicate and keep the workflow running smoothly.

Broadcast News Workflow

News Channel 16 is a fictitious broadcast station in a medium-sized market. They are making the transition from tape-based production to digital broadcast using solid-state digital recording cameras and working to build an HD infrastructure for their future broadcasts.

The station is now using digital video cameras for their field crews. The station's old tape-based workflow does not translate to digital file-based processes and is being phased out, so a new workflow must be designed. The staff have had their initial evaluation meetings and developed their installation plan. During the evaluation process, the staff identified the following primary requirements for their new workflow:

- Keep track of the digital media files brought in from field crews after the breaking news packages are sent in via microwave transmission from the location shoot.

- Properly identify the current production stage of a package before it goes to air.
- Remotely verify the content of the package before it goes to air (producer review).
- Confirm legal releases before airing investigative pieces.
- Develop a near-term archiving system for media files after the news story has gone to air.
- Provide producers with a way to quickly browse HD material from their desktop PC computers in the newsroom.

After a lot of hard work and many marker board diagrams, they came to a consensus of how a story would track through the production process. The diagram here illustrates how their news stories flow through the four major phases of news production: ingest, edit, review, and playout.

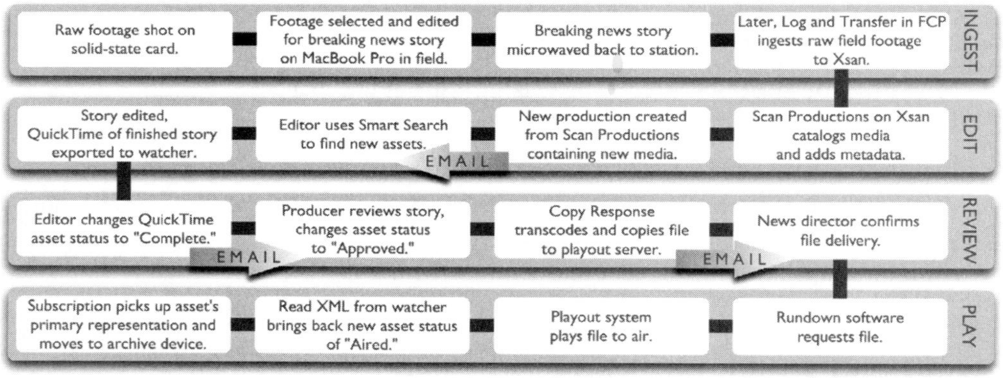
Raw footage shot on solid-state card.
Footage selected and edited for breaking news story on MacBook Pro in field.
Breaking news story microwaved back to station.
Later, Log and Transfer in FCP ingests raw field footage to Xsan.
INGEST
Story edited, QuickTime of finished story exported to watcher.
Editor uses Smart Search to find new assets.
EMAIL
New production created from Scan Productions containing new media.
Scan Productions on Xsan catalogs media and adds metadata.
EDIT
Editor changes QuickTime asset status to "Complete."
EMAIL
Producer reviews story, changes asset status to "Approved."
Copy Response transcodes and copies file to playout server.
EMAIL
News director confirms file delivery.
REVIEW
Subscription picks up asset's primary representation and moves to archive device.
Read XML from watcher brings back new asset status of "Aired."
Playout system plays file to air.
Rundown software requests file.
PLAY

As their workflow continues to evolve, they may come up with more issues or modifications. Final Cut Server's tools will allow for these changes any time the workflow requires them.

The News Channel 16 crew has gotten word of an alleged scandal in the city. The deputy mayor has been charged with embezzlement. As Channel 16 follows this story, you'll see how Final Cut Server meets the station's needs in the new workflow.

Uploading with Metadata

In the first example, you'll see how Final Cut Server's metadata tools can be used to track specific assets. We'll look at the assets that are created from the original files recorded on the solid-state memory cards. Since these unedited assets and the edited version of the news story (that was sent in via the microwave transmission) share common metadata, new processes introduced during the ingest process will help the staff distinguish the unedited assets from the edited footage sent in via microwave transmission.

The embezzlement story broke this morning. This afternoon, the accused official is holding a brief news conference to state his innocence. The News Channel 16 field crew is on location at City Hall. After shooting the press conference and editing the footage on a MacBook Pro, they send the edited story back to the station from the truck via microwave transmission for a live update on air. Once aired, the file used for playout is uploaded to Final Cut Server as a reference for related packages, and to create a shorter version for the station's website.

Later, the crew arrives back at the station with the original, unedited files on the memory cards. A field crewmember hands the cards to an assistant editor, who then converts the original files to QuickTime movies, using the Log and Transfer feature in Final Cut Pro. She adds the term *field_footage* as logging information in the Log and Transfer window. This information will be included as metadata of the assets when they are uploaded into Final Cut Server, saving editors' time when they search for these assets in the Final Cut Server catalog.

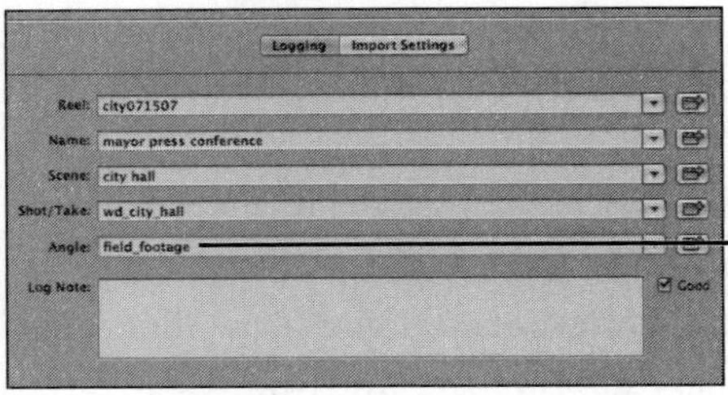

The assistant editor uses the Log and Transfer tool in Final Cut Pro to add information about field footage, which will appear as metadata in Final Cut Server.

At this point, the new Final Cut project that the assistant editor has created is being used only for the log and transfer process, and this is really all that's needed. Once all the raw media files have been transferred and the proper metadata has been assigned to each file, it's time to get everything into Final Cut Server.

The assistant editor saves the Final Cut Pro project, and then begins the upload by dragging it directly into the Final Cut Server interface and onto an existing production called Embezzlement. This production was created when the story broke, and it will act as a container for all related assets added in the future.

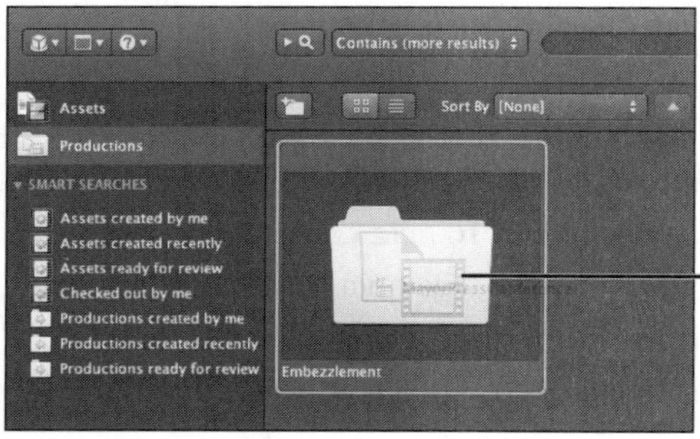

Dragging a Final Cut Pro project onto a production in Final Cut Server begins the upload process.

When uploading a Final Cut Pro project, the upload process brings along all the original unedited media. In a few moments, each media file is turned into an asset, and is also visible as an element in the new Final Cut Pro project asset and searchable in the catalog. The field_footage logging information is shown as metadata for the assets and can be used for searching by other Final Cut Server users.

> **NOTE ►** Project uploading is just one way to get media files from a log and transfer process into Final Cut Server. You could also create a frequently scheduled scan on the assistant editor's scratch disk location on an Xsan volume, which would add the newly transferred files as assets in Final Cut Server. But what about that specialized metadata? Using FCP 6.0.3 (or higher) and metadata mapping, users can include logging information from P2 and XDCAM files directly inside of logged and transferred QuickTime files. With a few custom metadata maps, the metadata from those QuickTime files could map into same-named metadata fields within Final Cut Server. See the Knowledge Base article on metadata mapping, which you can find on Apple's Final Cut Server support website, for more details. This feature will need to be set up by your administrator.

The uploaded assets now have metadata that came from the Final Cut Pro log and transfer process. To further identify each asset in the Final Cut Server catalog, the assistant editor can add specific keywords in the asset's info window. These keywords can help users sort these new assets with other assets that may already be in the Final Cut Server catalog for future news stories.

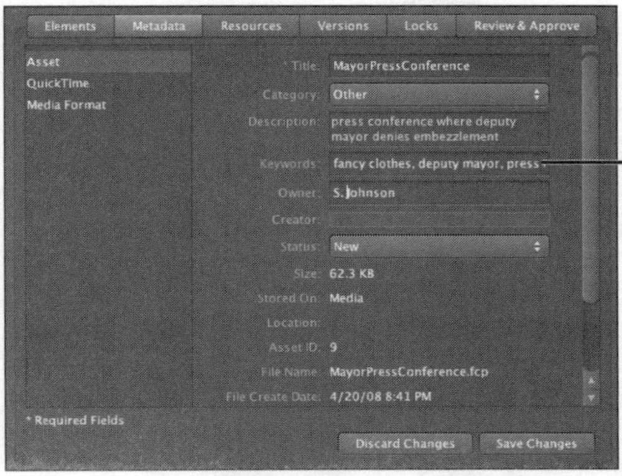

Double-clicking an asset in the production window will open its info window, where keywords can be added.

To find all footage on the scandal, the News Channel 16 editors and producers pull up the Embezzlement production. In the search field at the top of the Productions pane, they type *Field Footage* and find all the assets that share that metadata. Now they know they are looking at the original unedited assets.

They can browse through the assets in this search and quickly see which assets have the keyword *fancy clothes*. These original unedited assets are what they need to use in another news story package about how much the deputy mayor spends on his clothes.

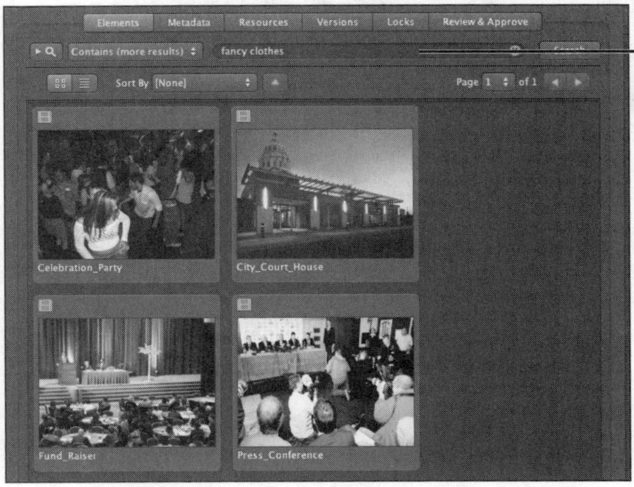

The search field at the top of the Productions pane is used to sort through all the assets in the production to find only the assets with *fancy clothes* as a keyword.

Tracking Production Progress

The scandal continues. The News Channel 16 crew has been adding to the Embezzlement production. Many media assets are available in the station's catalog showing the deputy mayor stating his innocence, as well as more recent material showing him at a recent city gathering in expensive clothing.

Today is the deputy mayor's arraignment. At any moment, footage from the courthouse will arrive at the station and be added to the story for the 6 p.m. broadcast. The news director needs to know exactly when this story is complete and make sure the file is in the playout server for the rundown to play out. When the Embezzlement production was first created, the Package metadata set was used to categorize the production, because it both describes the end deliverable and contains the very useful Status metadata field for status tracking. At the beginning of this new news day, the news director sets the status of this production to New from the pop-up menu.

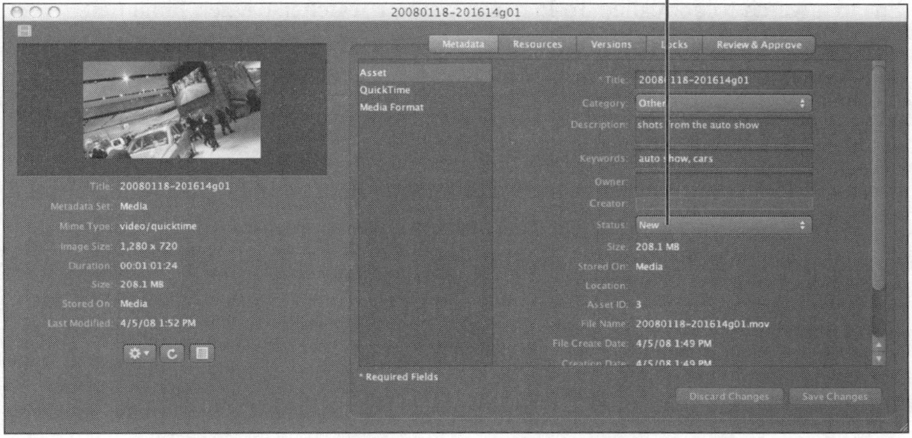

The craft editor then searches for the new assets in this production. He's got a lot of material to choose from.

He then checks out a Final Cut Pro project that was started by the morning shift editor: DepMayor Scandal—day2. He adds today's assets to this project by dragging them from the Final Cut Server window right into the Final Cut Pro Browser window.

NOTE ▶ The craft editor can directly drag and drop from Final Cut Server to his Browser in Final Cut Pro because these assets are on the Xsan volume, and his machine is an Xsan volume client.

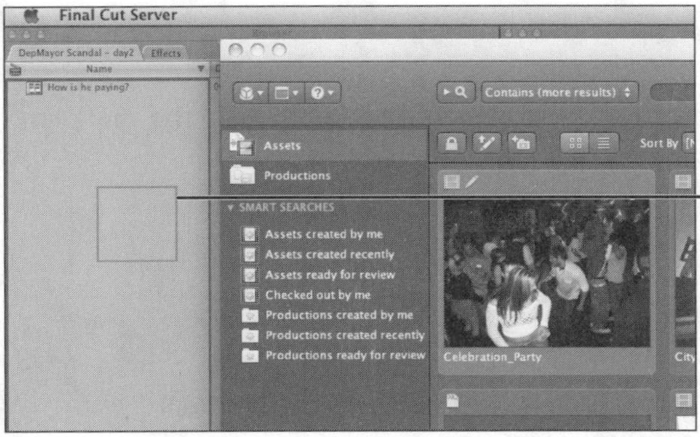

When the news package is finished, it then needs approval by the story's producer. The editor saves his project and exports the finished sequence as a QuickTime movie from Final Cut Pro to a watch folder called To Playout. Using a watcher trigger, Final Cut Server detects the file in the watch folder and fires off a copy response, which transcodes the file into the highly efficient H.264 codec and then transfers the file to the station's FTP server, just in case the producer cannot use Final Cut Server from the exotic locale she's at this week (see the next section). It then triggers yet another copy response, simply moving the file as is from the To Playout folder into the main Media folder of the Xsan volume. In the process, it creates an asset and assigns it to the Embezzlement production. The automation is still not done! It then fires off an email to the producer, telling her that this story is in need of approval before it can air. Finally, it deletes the original QuickTime movie from the To Playout folder.

MORE INFO ▸ For elaborate automations like the one above to work, all of the important devices at the station, such as the main editorial Xsan volume and FTP Server, were previously set up as devices so that Final Cut Server could interact with them.

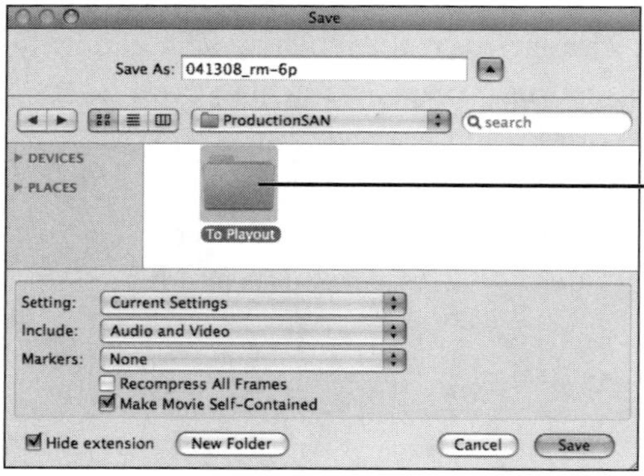

The craft editor saves the QuickTime movie into the To Playout watch folder, which will then fire off four responses.

The editor checks in the Final Cut Pro project, saving revision comments in the check-in window. He also changes the Status metadata field to Completed.

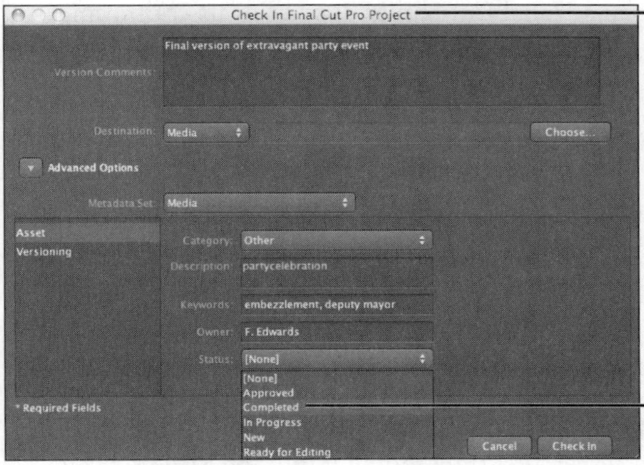

Choose Check In from the shortcut menu of the asset in Final Cut Server.

In the Check In Final Cut Pro Project window, the editor changes the story's status to Completed.

Getting Remote Producer Approval

One of the producers for the Embezzlement story is in Mexico, looking into odd trips the deputy mayor has been taking and whether any city business was conducted on the trips. She needs to approve her stories from her hotel. Because the Final Cut Server is "inside" the station's network, the remote producer needs to be able to connect to the station using a virtual private network (VPN) connection (set up by the facility's IT staff). Just in case she can't get in, very small versions of her stories are on a public-facing FTP site that she can view from a web browser, and then she can call in her approvals from her cell phone.

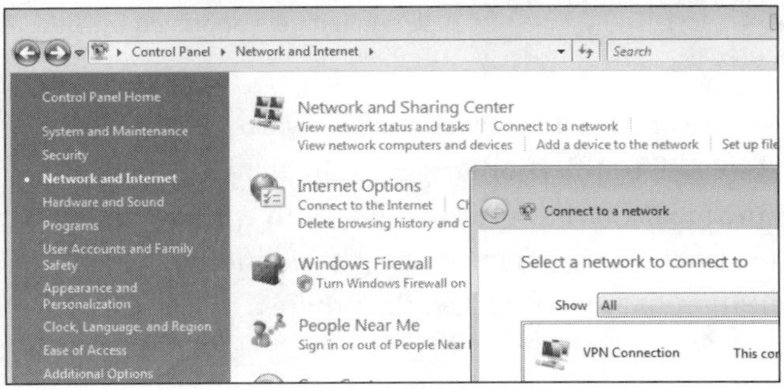

The producer receives her email notification in Mexico and connects to the Internet and the station's VPN from her hotel room. Once the VPN connection is made, she launches the Final Cut Server client software. Because Final Cut Server does all of its communication over standard network connections, if the network bandwidth is limited, she might experience slow performance.

> **NOTE ►** If the client software has not yet been installed, she can install it on the system by following the instructions in Lesson 1, "Overview and Interface Basics."

In the Embezzlement Productions pane, she selects the assets she needs to review and views a scaled-down proxy of each in QuickTime. If she agrees with the condition of the asset, she changes the status in the Review and Approve tab to Approved and saves the changes.

As soon as she changes the status to Approved, a subscription fires off a copy response, which transfers the movie to the playout server, first transcoding it into the MPEG-2 transport stream that the playout server needs to play the file when it comes time to air. An email is also fired off, telling the news director that the producer has approved the story.

Control-click the asset and select View > Proxy from the shortcut menu. A scaled-down version will play in QuickTime.

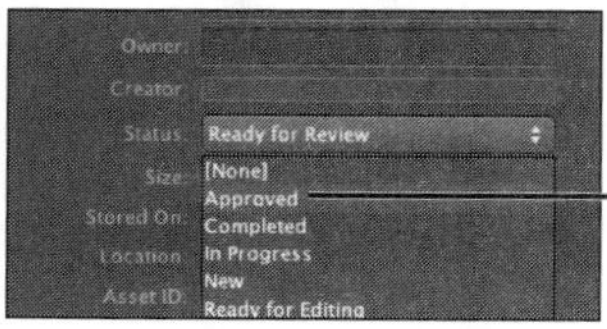

The asset is reviewed, and if it's up to snuff, it gets approved by the producer.

It's nearing broadcast time. The news director needs to track the story progress to make sure it's completed. After receiving the email notification, triggered by the producer's approval of the story, he also opens the Final Cut Pro project asset's info window and checks its status. Seeing its Completed label, he knows the editor has no more revisions to the story. Three people in the organization have now tracked the progress of a story without having to speak to each other.

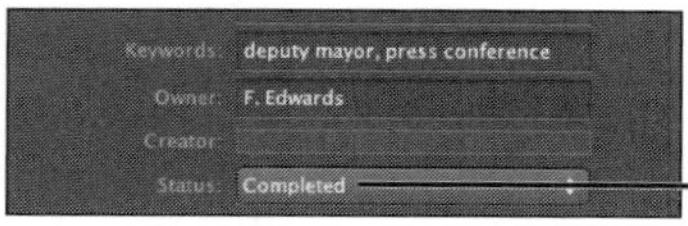

The news director sees that the editor has no more revisions to the story.

Now he needs to check the progress of the actual files for the evening broadcast. In Final Cut Server, he chooses Search All Jobs from the Server pop-up menu in the upper left of the client application window.

In the Search All Jobs window, the news director quickly verifies that the files have been transferred to the playout server. He can double-click a specific job to get even more details on its progress. If the news director were a more passive fellow, he might ask the Final Cut Server administrator to create a subscription to any job that had a Status equal to "FAIL" and Stored On equal to "PLAYOUT." This subscription could then send the news director an email notification, warning him of the error.

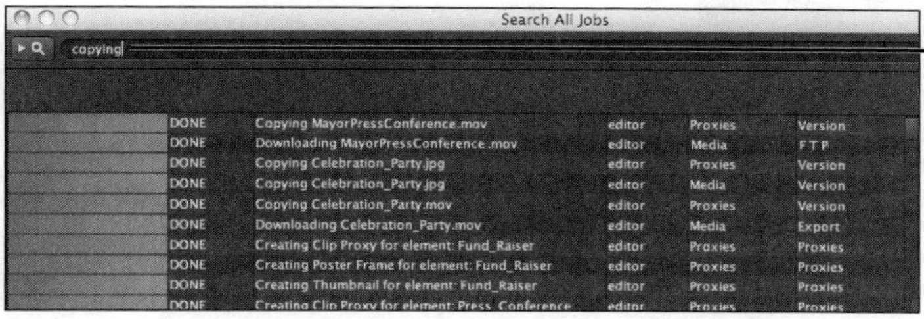

The news director quickly sees all the files that have been uploaded to the playout server by using the search field at the top of the Search All Jobs window.

Making Assets Available Based on Metadata

The Investigative Team (I-Team) at News Channel 16 has recorded an exclusive interview with a senior department manager at City Hall. No one else in the facility can view it until the releases are in place and the legal department has approved the story.

The I-Team staff will use the special controls in Final Cut Server to automate a process based on a specific metadata tag for these assets. Because these assets are sensitive in nature, they are stored on a specific Final Cut Server device with protected access for I-Team users only. An Xsan folder can be used to allow only specific group members to have access to the files through the Finder. Final Cut Server will use this same group information to control access to the assets and devices.

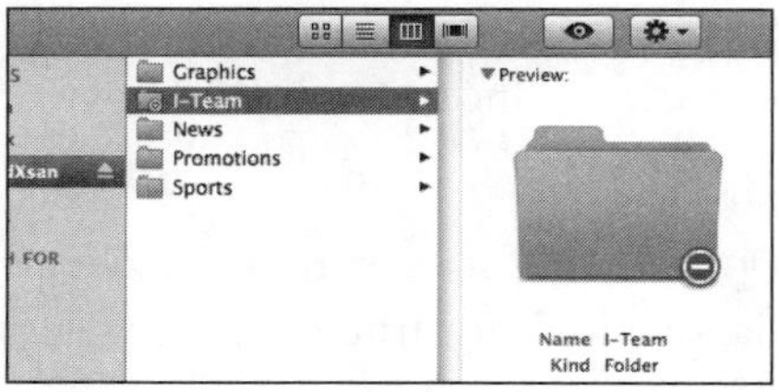

The legal department gives word that the interview is OK for release. The I-Team editor launches Final Cut Server and finds the interview footage in the main window. In the asset info window, she changes the lookup in the custom metadata field called Full Access Approved to Yes. Now a subscription to this metadata field triggers a copy response that transfers the finished file to a different device, where other users will be able to access the asset. This same trigger also sends an email to the editors informing them that the asset is ready for use in a news story.

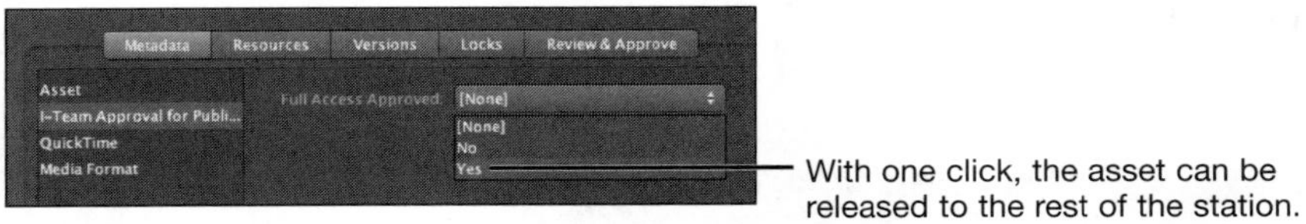

MORE INFO ▶ To learn how to build custom metadata fields and automations, see Lesson 8, "Advanced Administration."

Archiving Assets

The scandal has come to an end. The deputy mayor has resigned from his position. Now all the primary representations of any associated assets for this story can be moved from the primary storage device to an archive device, so that the next major stories that the station is working on have room to grow. Asset proxies and thumbnails remain where they are, to aid in searching, just in case they need to be restored down the line.

To perform the archive, a trusted staff member finds all associated assets by opening the Embezzlement production. The staff member then selects all assets within the production, and then Control-clicks and selects Archive from the shortcut menu.

Previewing and Annotating HD Footage Through Proxies

The producers at the station are creating a sample broadcast to test new HD content. Browsing through HD footage takes too much time and resources on an editing system. The proxy-viewing capabilities of Final Cut Server enable the producers to view lower-resolution versions on their own Windows-based PCs.

During the uploading process, a lower-resolution proxy is created for each asset. Proxies can be generated in a number of formats to meet the bandwidth infrastructure at the television station. If a lot of people will be viewing these files, the administrator might want to customize the proxy format for the station so that it takes up as little bandwidth as possible while still offering the quality needed for thoroughly reviewing the content.

The producer sits down at her desk and launches the Final Cut Server client application. Even though her computer is a PC running Windows, the Final Cut Server interface and features are exactly the same as in the Mac version. She uses a previously created Smart Search on the left side of the main interface to find all the HD assets. The info window for each asset displays all the metadata information about the asset. As she clicks the Proxy Preview icon for each asset and watches the clips, she decides to add comments, so she opens the annotations window.

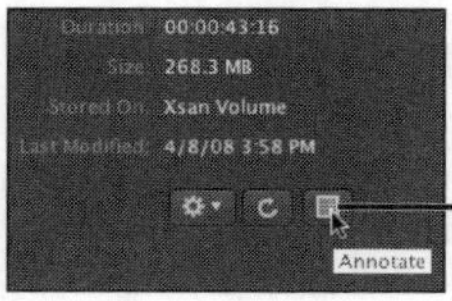

In the info window, click the Annotate button to see a preview window and add comments.

In this window she previews the entire HD asset in low-resolution proxy form. She uses the Mark In and Mark Out buttons under the preview viewer to select a portion of the asset for annotating, and she enters her comments.

Annotations for Island Getaway
Annotation
In
Out
This is a great shot. Need closer footage of fountain. Keep closeup on the tree at this point
00:00:01;16
- JonR on Apr 27, 2008 7:48:54 PM
00:00:07;02
Annotation:
This is a great shot. Need closer footage of fountain. Keep closeup on the tree at this point

She clicks the Open Asset View icon to return to the info window for the asset.

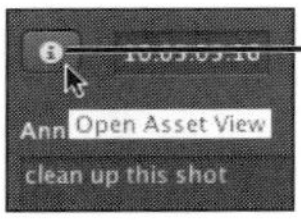
Open Asset View
clean up this shot
Click here to jump back to the asset's info window.

Now that we've seen how Final Cut Server can touch every aspect of the workflow at a hectic television station, let's see how other tools can be used to aid the content-creation process at a much different organization.

Rich-Media Publishing Workflow

C3 (Central City Communications) is a "new model" publishing company creating feature-based multimedia content for distribution in various ways. Their staff includes creative photographers who capture rich, exciting photos and video for the publication. Their writing and editorial staffs use an efficient collaborative process for creating text in multiple languages. They have an established new-media workflow that already provides many advantages over the older print-based models. There is a desire to further streamline some of their processes and ensure that their workflow is dynamic enough to respond to changes as opportunities present themselves. The staff of C3 is experienced in file management and wants to use the more advanced features in Final Cut Server.

We will follow the process of creating a standard rich-media feature package as it works its way through the organization. We will see how the existing workflow processes have been maintained in the Final Cut Server deployment, and we'll examine new capabilities and efficiencies that have been added by Final Cut Server's advanced features.

C3 has embarked on a long-term project to provide feature-rich multimedia content before and during the grand opening of the new International Art Museum in Shanghai, China. Their photographers are already in Shanghai gathering initial images and video content. Writers have started the editorial process with stories about the local cultures and destinations.

This is the most ambitious venture the group has ever undertaken. While their existing digital workflow has worked perfectly for normal feature projects, they have identified the following new requirements for the Shanghai project:

- Moving many kinds of files in many formats to and from a large, centralized catalog. Formats include Word documents, images, and video files. They also need the ability to add content remotely and have it tracked dynamically throughout the entire workflow.
- Tracking specific versions of files for the editorial process.
- Automating email notification, copying files to a website via FTP, and other responses based on specific triggers in metadata.

Remote Workflow

The C3 team of digital photographers, which is based in New York City, is on location in Shanghai. They need to add their images to the Final Cut Server catalog from their hotel rooms at the end of the day, saving overnight shipping costs to the United States, and also make it easy for the NYC staff to access and track these assets quickly through the use of metadata. We will see how the team members can access the Final Cut Server catalog remotely via a VPN connection and upload new assets with specific metadata tags.

After a day of shooting at a construction site for the new art museum, the digital photographer returns to his hotel room. He downloads the images from his digital still camera to his laptop and reviews them. After choosing the images he wants to use, he connects to the Internet, activates a VPN session to the C3 intranet, launches the Final Cut Server application, and logs in. He uploads the files by dragging them from their laptop locations into the existing Shanghai Art Museum production. The Upload window appears, where he adds *local artwork* to the keywords metadata field along with any additional metadata. He and other users will continue adding assets to this production during the remote trip, keeping all the assets related to each other.

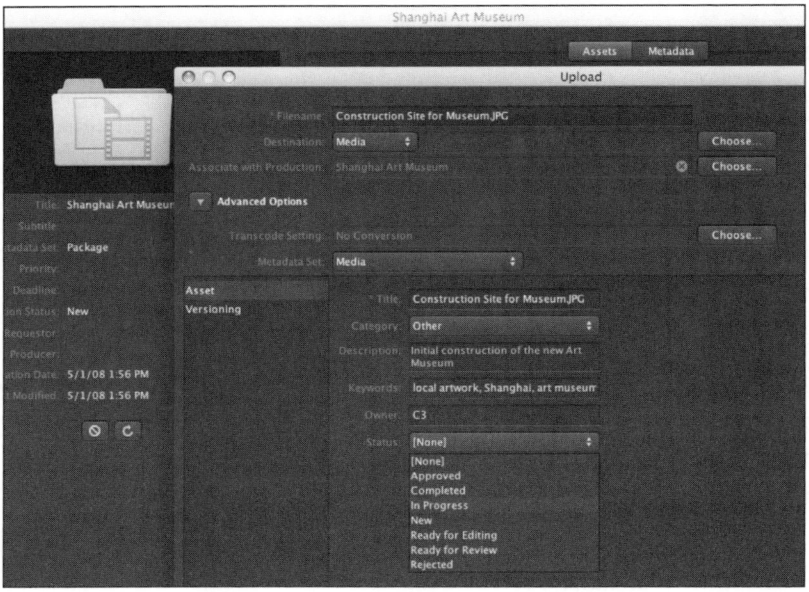

Accessing and Adding Metadata

Evening in Shanghai is the same day's morning in New York City. The C3 team members in the home office need to quickly find specific assets from the photographers for use in a piece on fashion.

Using the search tools in Final Cut server, the C3 staff can quickly find specific assets in the catalog. We will see how using Smart Searches saves the team time and improves the overall efficiency of the process.

In the main window of Final Cut Server, a NYC editor uses the advanced search features to look for assets with the keyword *local artwork*, which the Shanghai photographers added earlier. All assets with this keyword tag will appear quickly and be ready to use for the fashion piece. The editor saves the list of sorted assets as a Smart Search, naming it *Local Chinese Artwork.*

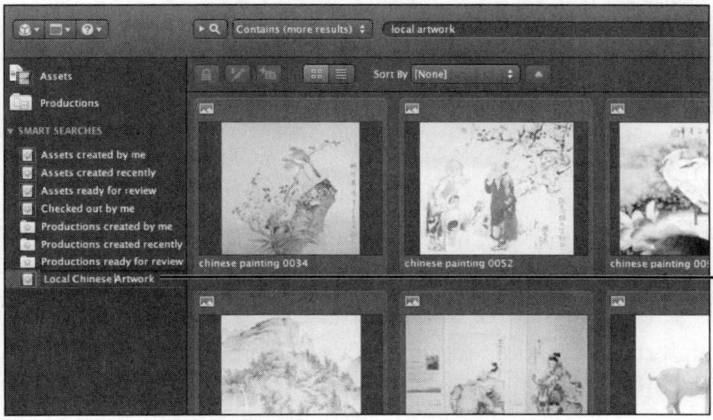

When the desired search results are displayed, click the Save as Smart Search button in the bottom-right corner and name the Smart Search here.

Versioning Assets

The organization's current versioning process works when the writers and editors are down the hall from each other in the New York office. For the Shanghai project, the team needs a versioning process that will ensure no information gets lost during the process and all changes can be tracked. This is especially true of Word documents as they go through the editing process.

Final Cut Server's version control provides a record of users and their changes. Final Cut Server will store the versions it creates of each asset in the version contentbase, which is created during initial installation of Final Cut Server by default.

An important decision to make is the maximum number of versions that can be associated with an asset. Once the limit is reached, the oldest versions will be deleted as new versions are added.

With version control in place, staff members in the NYC office can track different versions of their co-worker's Word documents as they continue to make editorial changes. Anyone can access these versions by double-clicking on the Word Document asset and clicking on the Versions tab. These versions were created each time the asset was checked in; editors can revert back to older versions if needed.

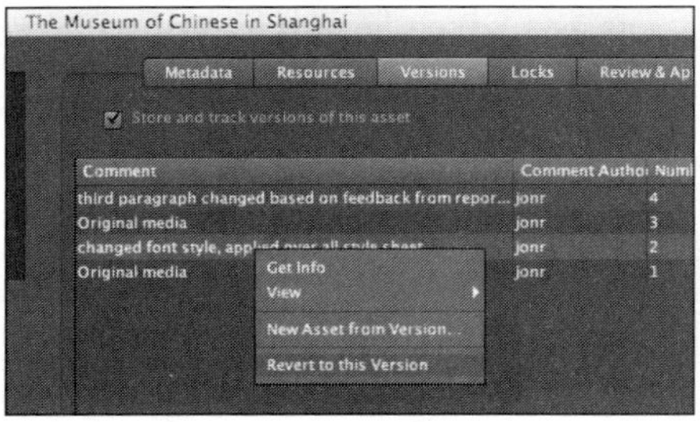

NOTE ▶ Enabling version control will make Final Cut Server create specific copies, or versions, for any asset that is uploaded to Final Cut Server. This has implications for the amount of space you have on your storage devices, so use this feature only when you are sure you have enough room on your storage devices. For more on version control and contentbases, see the Final Cut Server Setup and Administration Guide.

Automating Posts to a Website

The staff needs to have custom-built automations for their workflow, which will streamline their process by reducing repetitive, manual steps. This includes email notification and FTP transfers of assets at the completion of a production process.

One of the first feature segments for the Shanghai project is complete, and it is now time for a staff member to transfer it to the company website. In the past, he would use a standard FTP software program on his computer to manually "push" the file up to the web server. This is not an efficient process when there are a great number of files to transfer, and they need to go to specific directories on the web server. A custom automation will not only simplify the transfer process but also eliminate the possibility of sending the file to the wrong directory on the server.

After consultation with the user, the administrator decides to create a custom automation, which can be used for this and all future completed segments. Two things have to happen: Final Cut Server transfers the file to the FTP web server, and then it sends an email to the webmaster notifying her of the new file.

In the Administration Panel, he creates two responses with exactly these actions:

- The copy response is set up to copy the asset to the web server, and then to a specific folder within the web server.
- The email response can be customized to include metadata from the assets.

Finally, he adds these responses to a subscription, which will be the trigger. The responses will fire when the status metadata tag is changed to *Completed*.

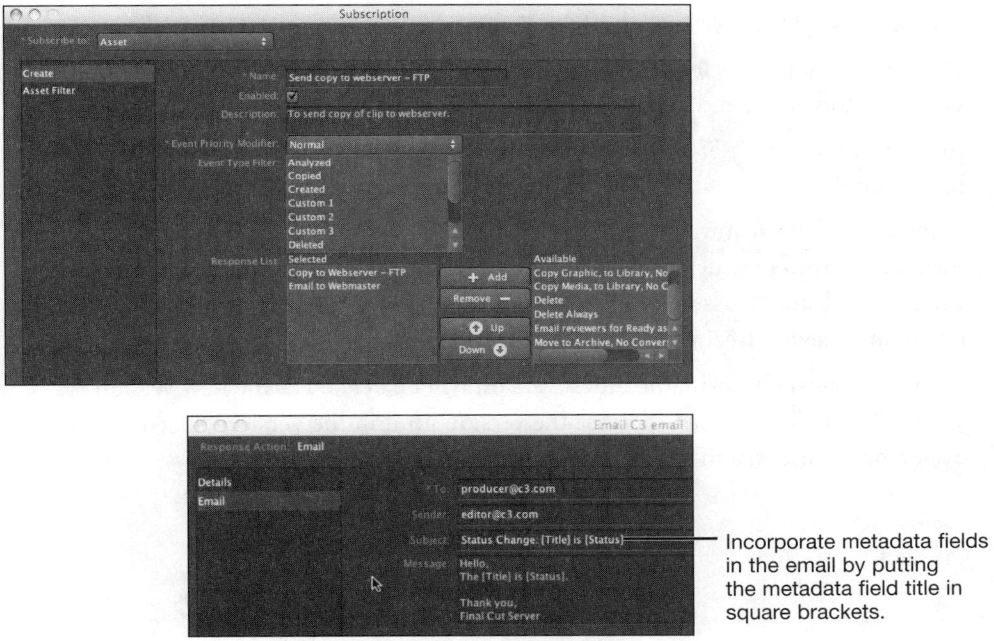

Email responses can be customized in the Administration window.

Lesson Review

1. How can producers use the check in/check out features of Final Cut Server if they are in a remote location and *not* using an edit-in-place workflow?
2. How would producers be able to comment on recently uploaded assets?

3. Where would an editor change the current status of a production from Work in Progress to Completed?
4. How would a post-production house free up space on a main device?
5. If the administrator enables version control within Final Cut Server, what happens when the limit of versions has been reached?

Answers

1. They need an Internet connection to connect to the organization via a VPN (virtual private network). They then download the Final Cut Server client application and log in.
2. Using the annotations feature within Final Cut Server, select certain portions of the video asset using In and Out points and make comments to these selections.
3. In the production's Metadata tab under Status. There are a variety of lookup choices here, including Ready for Review, Hold, and Approved.
4. Using the archive feature in Final Cut Server. A post-production house can move big media files from the current device to an archive device. When the archive feature is used, it will leave an asset in the catalog but move its large primary representation to the archive device, freeing up space on the main device.
5. During the next check-in, the oldest version will be deleted as the new version is added. Administrators can increase the version limit in the Version Control tab of the Preferences pane within the Administration Panel.

Appendix

Best Practices for Installation

This appendix serves to prepare you for the installation of Final Cut Server at your organization, as well as to guide you through the installation of the software.

If you've already installed the software, you will also find information that can help you optimize its performance.

Final Cut Server has the potential to greatly tax the infrastructure of your organization, especially if it is engaged in the following tasks:

- Searching a large catalog (50,000+ assets) by many users simultaneously
- Performing *full* scans (as opposed to add-only scans) of a very large volume (such as an Xsan volume) during a time of day when the volume is being heavily used by the facility, or if the volume's contents change significantly from day to day
- Transcoding files during copy operations on a regular basis, especially if those transcode requests involve converting files into or out of GOP-structured codecs (for example, MPEG-1, MPEG-2, MPEG-4, H.264, HDV)

To guard against the bogging down of your infrastructure, take care to ensure that the design and component architecture of both your Final Cut Server and your organization's network infrastructure are able to handle the load that Final Cut Server software will place upon them.

We'll first delve into ideas for optimal environmental conditions for the Final Cut Server. The appendix concludes with comprehensive steps for installing Final Cut Server software.

Optimizing the Final Cut Server

The following recommendations are specifically for the Mac that will run the Final Cut Server software. Apple's minimum and recommended requirements for the Mac that runs Final Cut Server are outlined in the "Before You Install Final Cut Server" document found on the Final Cut Server installation CD. In this section, we will go further into best practices to ensure even greater reliability and performance.

Using the Right Mac

The Mac running Final Cut Server should be one of Apple's latest offerings. Most preferable is an Intel-based Xserve. The raw processing power, field-replaceable components, redundant power supplies, and hard drive modules are reason enough to use it as the Final Cut Server.

Stuffing the Machine with RAM, RAM, and More RAM

The Apple-recommended 4 GB of RAM should be considered just a starting point for Final Cut Servers. There are two rules of thumb:

- 1 GB of RAM per core in the server (for example, at least 8 GB of RAM for an 8-core Intel Xserve)
- 8 GB of RAM for every 250,000 assets that will be in the catalog, up to the maximum 32 GB that can be placed in an Intel Xserve or Mac Pro

Creating a Redundant Array for the System Volume

Because the Final Cut Server will continuously gain significance as a key component in the workflow of your facility, you should consider creating a RAID 1 array out of two hard drive modules and installing the OS and Final Cut Server on top of that array. In this configuration, one drive can completely fail without the system going down. When the failed drive is replaced, the reconstructed RAID array can provide redundancy once again. In addition, if the Xserve has a spare drive, you can clone a backup copy of your RAID array on that drive. In case you have system corruption, you can restore the data from it and get back online in a hurry.

> **NOTE ►** This configuration refers specifically to running Final Cut Server. Your media catalog should be located on separate drives and backed up regularly. See "Planning File Locations" in this appendix for more detail.

Beefing Up the Ethernet Pipe

Another strongly recommended component is the addition of a multiport Ethernet card for the Intel Xserve or Mac Pro. This allows you to use Link Aggregation Control Protocol (LACP), which is built into Mac OS X, to combine these ports into a single addressable "pipe" for access to the server. Doing so ensures that all users can get the bandwidth they need when asking for any cached or exported file through the Ethernet network. Having the extra card allows the built-in Ethernet ports to do their usual job: general administration of the server and/or connection to the metadata network of an Xsan system.

Apple sells a two-port PCI Express card, and third-party vendors, most notably Small Tree Communications (www.small-tree.com), also sell multiport cards that serve the same function.

A required complement to the multiport Ethernet card is to have a switch on your organization's LAN that supports the LACP protocol. LACP works only if both the computer and the switch can talk this protocol. You can find more information about LACP by searching on the Internet for the term *IEEE 802.3ad* (the technical specification for the protocol, standardized by the Institute of Electrical and Electronics Engineers).

Regulating Bandwidth for Xsan Clients

A Final Cut Server that is a client of an Xsan system brings one certain truth: During any copy or transcode operation that involves files on the Xsan volume, which includes the creation of proxies during uploads and scans, the Final Cut Server will pull a tremendous amount of bandwidth availability away from the other clients.

There are two ways to "govern" this bandwidth pull:

- Create a single path of Fibre Channel cable from the Final Cut Server to the Xsan system's Fibre Channel switch. Further, you could configure that path to negotiate down to 1 Gb (yes, that's *gigabit*, little *b*), rather than the normal 4 or 2 Gb. A 1 Gb connection would pull only a maximum of about 100 MB per second. The downside of this method is that with a single connection, you have a single point of failure in that one cable.
- Use Vmeter SQM (SAN Quality of Service Manager) from Vicom Systems (www.vicom.com). Vmeter QMS is a simple program that sits on top of your Fibre Channel card's driver and regulates its bandwidth to a certain maximum.

> **NOTE ►** It should go without saying—but for the tinkerers out there, let's be clear—that a Final Cut Server cannot also be an Xsan metadata controller. If it "lives" on an Xsan, it has to be a client of that Xsan, and leave the important processes of running the volume to a pair of additional Macs that are exclusively the MDCs.

Optimizing Your LAN

Besides having a switch on your LAN that can speak LACP protocol, as mentioned earlier, you must also ensure that the LAN can handle the intensive traffic that Final Cut Server will provide.

If you don't have Gigabit Ethernet switches at your facility, it's time to upgrade to this speed. The costs for these switches are at or below those of 100Base speed switches sold just a few years ago, and the difference in performance will be dramatic.

DNS, or Domain Name Service, is also a critical component if your Final Cut Server will serve a user base of 20 or more. One benefit of having a DNS locator for the Final

Cut Server is that users will be able to type its name into their web browsers when first downloading the client application, as mentioned at the end of Lesson 1, "Overview and Interface Basics."

For Xsan systems, it's very important to have well-established public and private (or metadata) Ethernet networks. Make sure, through the Xsan Admin program, that your metadata network is designated to run on the infrastructure that you intended. Final Cut Server traffic erroneously traveling on the metadata network of an Xsan will grind performance to a halt.

If you are interested in having users literally outside of your walls access the Final Cut Server, then the placement of a VPN firewall appliance between your LAN and the ISP router of your Internet service provider (ISP) is required. There are many choices out there, for small and large facilities alike.

Planning File Locations

Final Cut Server interacts regularly with the storage systems, or devices, at your facility. Because of this, thoughtful design of these devices will go a long way toward ensuring that the system works well.

Many devices at your organization may pre-date your Final Cut Server, but most of these will probably be either network file servers or Xsan volumes, which hopefully have provided the throughput needed by your organization up to this point.

However, some devices will be created specifically for the Final Cut Server, and it is these devices that we will concentrate on. Well-designed devices that serve the needs of the Final Cut Server will greatly enhance the system's ability to serve your users.

Planning for the Proxies Device

As explained in Lesson 8, "Advanced Administration," the *proxies* device is a contentbase-type device in which Final Cut Server stores all of the clip proxies, image poster frames, and thumbnails for all the assets within the catalog. It is created during the installation process. Because of this, its location and configuration must be planned *before* installation, in the same way that you prepare the baby's room before the delivery date.

Your first task is to estimate the size of the proxies device. If you use the default H.264-based settings for clip proxy creation, you should estimate approximately *3 GB of hard drive space for every hour of content* that Final Cut Server will have in its catalog. This formula is derived from the 800 to 860 KB per second that these H.264 clips use, which varies due to frame rate and aspect ratio. All proxies play in this rate range, regardless of the codec of their primary representation. Plenty of additional free space will ensure that your organization can grow into the device. It's a difficult exercise, but try to estimate your facility's storage needs in three to four years and allocate that much space to the proxies device. You will avoid going above the 80% full range, which is where all storage devices begin to lose performance eventually.

Next, you need to consider how the proxies device will connect to the Final Cut Server. Since clip proxies, poster frames, and, most important, thumbnails are continuously being delivered by Final Cut Server to the users, this storage needs tremendous speed. Because of this, you should use storage that is directly attached or an Xsan volume. It also helps for the system to be redundant in case of drive failure. The following scenarios illustrate some ideas for optimal connection, the first being ideal:

- The proxies device is located on an exclusive direct-attached, hardware-controlled, RAID 5 array, using Fibre Channel, SATA, SAS, or SCSI protocols. Please note that USB or FireWire drives would not be part of this scenario!
- The proxies device is located on a RAID 5 array created by using an internal Apple hardware controller card and the drive modules located within an Xserve or Mac Pro.
- The proxies device is located on a unique affinity of an Xsan volume. This means that a unique storage pool would be created, and that a folder on the volume would be created for the proxies device that would have an affinity pointing to this unique storage pool. An Access Control List (ACL) should then be placed on this folder to restrict access for staff members so that they don't accidentally move or delete the bundle file that contains the proxies device.

  Remember that Xsan volumes that pre-date the Final Cut Server can have additional storage pools added without the volume having to be reinitialized. Because of this, any organization can provide the unique storage pool needed for proxies regardless of the age, size, or condition of their Xsan volume.

In any case, the proxies device should *never* be placed on the system drive of the Final Cut Server itself. Doing so would significantly reduce performance of the system.

Planning for Other Final Cut Server Devices

During the installation process, in addition to the proxies device location, Final Cut Server will ask for the Production Media location. Within this location, Final Cut Server will create any number of additional devices specific to your customer profile.

> **MORE INFO ►** See the Setup and Administration Guide for details on all customer profiles.

For example, if you select the Video Production customer profile during installation, Final Cut Server creates three folders within the Production Media location: Library, Media, and Watchers. It then makes each of these three folders a device. The Library device is intended as a place to store stock files needed by the organization on a regular basis. The Media device stores media files. Finally, the Watchers device is where Final Cut Server places two watch folders, Graphic and Media. It then attaches watcher automations to these folders (see Lesson 8, "Advanced Adminstration"). Watchers will function best if they are being shared through traditional file sharing protocols, such as AFP, SMB, or FTP. Be sure you can get the help you need to fire those services up on either the Final Cut Server or the server that will provide these functions.

What's important to keep in mind, therefore, is that the location you pick for the source of these devices should also be on a fast and redundant storage system, either directly attached to the Final Cut Server or on an Xsan volume. As opposed to those of the proxies device, however, the size and speed requirements of this location are largely dependent on how the organization will use them. If the files on these devices need to be directly accessed by other Xsan-attached computers, then placing the Production Media location on the Xsan volume is a no-brainer.

If you turn on Version Control and Edit Proxies during installation, Final Cut Server will create contentbases for these features in the Production Media location. Because of this, you might want to consider restricting access to this location to users who might be able to access it by traditional means, such as file sharing or Xsan-attached machines that will be able to "see" the location.

Planning the Catalog Backup Location

Something that you have time to plan for, since it is not required during installation, is the location of the catalog backup files. As explained in Lesson 7, "Basic Administration,"

Final Cut Server has a built-in feature to back up its catalog (but not the original media files or proxies) to a specified location on a regular basis. The storage system used to store these backup files should meet the following criteria:

- The location should *not* be the system drive of the Mac, because that's where the catalog is.
- The location should be on a redundant array of hard drives, such as RAID 1 or RAID 5, if possible.
- The location need not be that fast, since the catalog backup can take as long as it needs to and usually happens during off-hours. FireWire drives, especially ones that have built-in RAID arrays, are a fine choice for the catalog backup location.
- The location needs to have enough capacity to accept each backup. If you are backing up daily, this will rapidly take up space if your catalog is large. A rule of thumb here is to plan for 1 GB of space for every 50,000 assets in your catalog.

Optimizing Transcoding Operations

If you anticipate the heavy usage of the transcode function with Final Cut Server, consider utilizing a Compressor Cluster rather than the single instance of Compressor that installs with Final Cut Server. A Compressor Cluster is a series of computers running Compressor in tandem, tackling a complex transcoding job by splitting it among themselves and then putting the completed file together by handing their individual contributions back to a central machine, called the Cluster Controller, for final assembly.

For example, if your organization regularly accepts files in many different formats and intends to use Final Cut Server as an equalizer for those files, transcoding them as they copy or upload onto your devices, a Compressor Cluster would be ideal.

Any organization that has or will have 250,000 or more assets should also plan to have a Compressor Cluster handle their transcode operations.

When creating a Compressor Cluster, each Mac will need to have a license of either Final Cut Studio or Final Cut Server, so that each instance of Compressor can run concurrently. Full instructions on how to create a Compressor Cluster can be found in the Documentation folder of your Final Cut Server installation CD.

Installing Final Cut Server

Before you begin to install the software on the Mac that you intend to be the Final Cut Server, make sure you have the following prepared:

- Software serial number.
- Customer profile you wish to use. Further details about customer profiles can be found in the Setup and Administration Guide that came with the Final Cut Server software.
- Locations for both your proxies device and Production Media. These will both be requested during installation. For more information about these locations, see the "Planning File Locations" section earlier in this appendix.
- Domain name or IP address of your non-authenticated SMTP (outgoing mail) server. You might have to call your ISP or IT department to get this.
- A display, if you're installing Final Cut Server on an Xserve. Even if that Xserve will not usually have a display connected to it, you must temporarily connect a display in order to wake up its graphics card. Only then will the installer recognize the graphics card and allow you to install the software. After installation, you can disconnect the display.

1 Insert the Final Cut Server CD into the Mac that you intend to be the Final Cut Server.

The CD window appears.

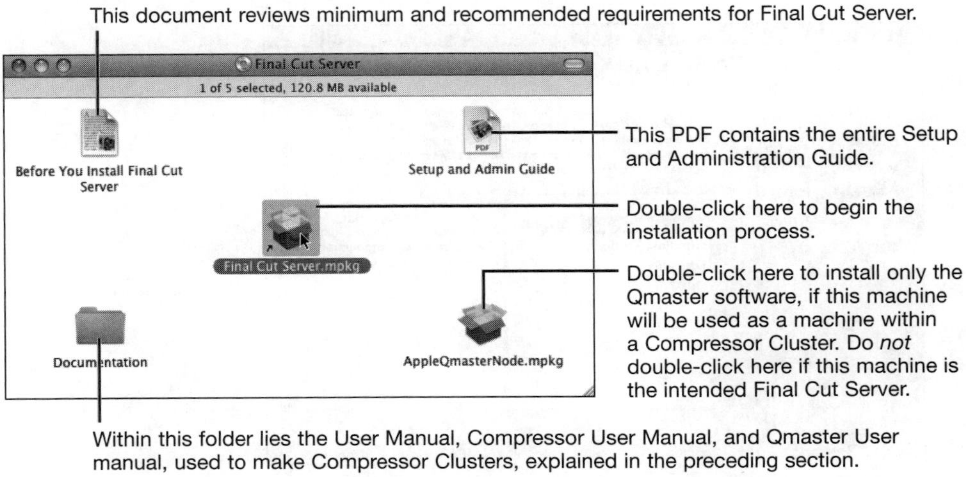
This document reviews minimum and recommended requirements for Final Cut Server.
Final Cut Server
1 of 5 selected, 120.8 MB available
Before You Install Final Cut Server
Setup and Admin Guide
Final Cut Server.mpkg
Documentation
AppleQmasterNode.mpkg
This PDF contains the entire Setup and Administration Guide.
Double-click here to begin the installation process.
Double-click here to install only the Qmaster software, if this machine will be used as a machine within a Compressor Cluster. Do *not* double-click here if this machine is the intended Final Cut Server.
Within this folder lies the User Manual, Compressor User Manual, and Qmaster User manual, used to make Compressor Clusters, explained in the preceding section.

2 Double-click the Final Cut Server.mpkg file to begin the installation process.

Read through the following few panes of introduction and the ever-present license agreement. You will then be prompted for your serial number.

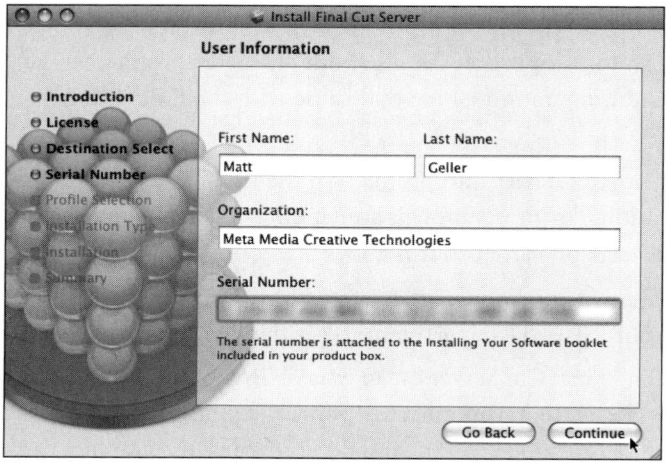

3 In the appropriate fields, enter your name, organization, and serial number, then click Continue.

The next pane will ask you for the customer profile you wish to use.

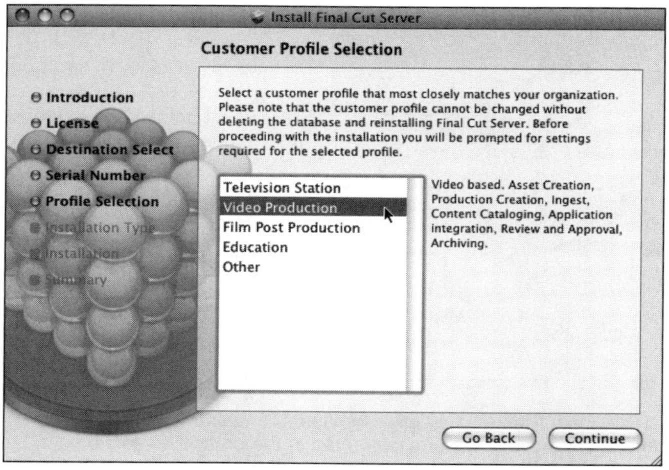

4 Select the appropriate customer profile and click Continue.

The next pane asks for storage locations and other important information.

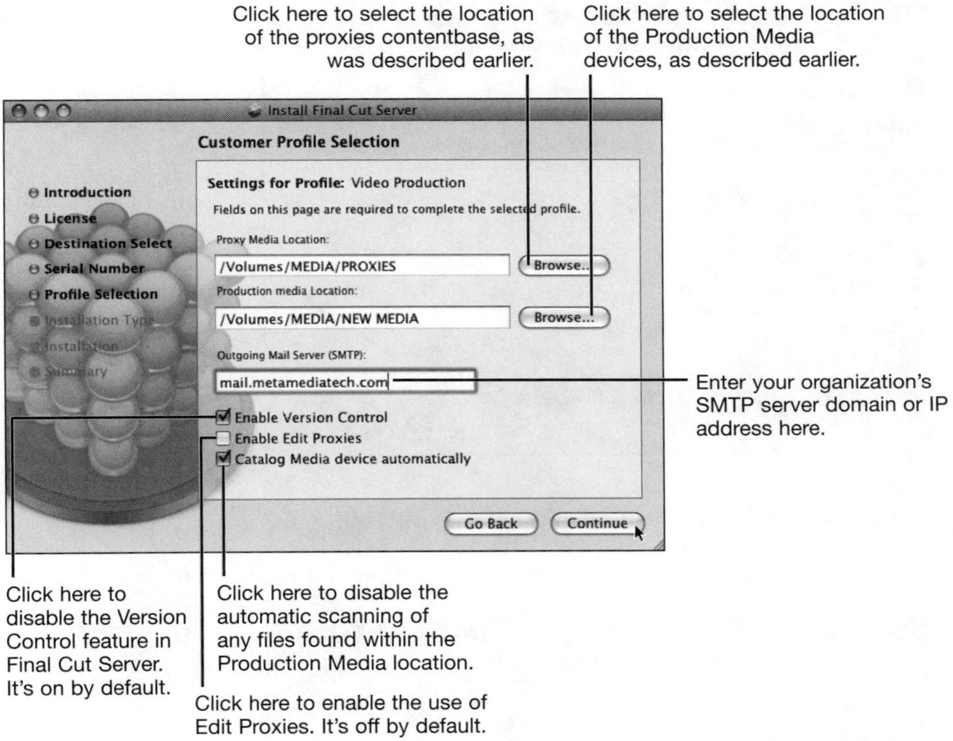
Click here to select the location of the proxies contentbase, as was described earlier.
Click here to select the location of the Production Media devices, as described earlier.
Install Final Cut Server
Customer Profile Selection
Introduction
License
Destination Select
Serial Number
Profile Selection
Installation Type
Installation
Summary
Settings for Profile: Video Production
Fields on this page are required to complete the selected profile.
Proxy Media Location:
/Volumes/MEDIA/PROXIES
Browse...
Production media Location:
/Volumes/MEDIA/NEW MEDIA
Browse...
Outgoing Mail Server (SMTP):
mail.metamediatech.com
Enter your organization's SMTP server domain or IP address here.
Enable Version Control
Enable Edit Proxies
Catalog Media device automatically
Go Back
Continue
Click here to disable the Version Control feature in Final Cut Server. It's on by default.
Click here to disable the automatic scanning of any files found within the Production Media location.
Click here to enable the use of Edit Proxies. It's off by default.

5 Fill out the sections and click Continue.

The next pane confirms your installation location.

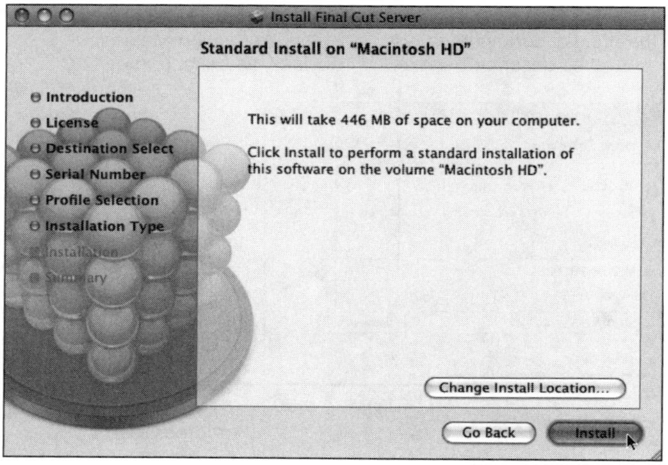

6 Click Install to begin the installation of the software.

When it's finished, another pane shows that the process is complete.

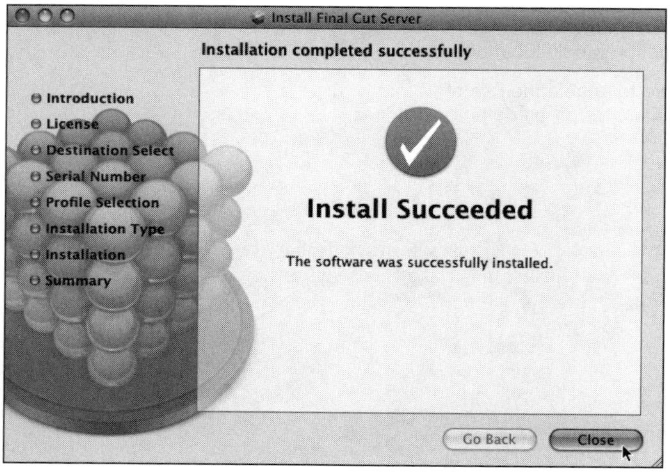

7 Click Close to quit the installer.

Now that your software is installed, see les 7 and 8 (in that order) to learn how to configure the other aspects of Final Cut Server to meet the needs of your organization.

Glossary

A

Add to Cache The process of downloading an asset's primary representation and placing it in the cache on your hard drive. This file then stays in the cache, and when you drag and drop the asset using Final Cut Server's interface, a link for that file's location is passed to the location or application to which you drag the file.

analyze A process, available only to administrators of Final Cut Server, that re-creates an asset's other representations, using the primary representation. This is usually done because the primary representation was changed outside of Final Cut Server, or because other representations were not created or are missing.

annotation Timecode-based comments on a video-clip asset that can be viewed by others in your organization. Annotations usually refer to particular sections of the clip, specified by In and Out points.

asset A listing within the Final Cut Server catalog for a file that exists within one of its devices. Assets refer to one or more representations. These include the primary representation and may also include a poster frame, clip proxy, edit proxy, and thumbnail.

C

catalog The entire database of Final Cut Server, specifically referring to all of the assets and productions that have been created.

check-in The process of returning an asset's primary representation to the Final Cut Server catalog after being checked out and then modified externally. If version control is enabled for the asset, comments on the new version are recorded by the person checking the asset back in, and the old version of the asset is saved in the Version device. Once checked in, the asset's lock is removed and others may modify its metadata or check it out.

checkout The process of both locking and exporting an asset's primary representation for the purposes of external modification. While an asset is checked out, other users are prohibited from modifying the asset's metadata or other representations until the asset is checked in.

clip proxy A video clip that Final Cut Server makes from the primary representation of a video asset. It is much smaller and of lower quality than the primary representation, and as a result, it is easier to transmit over Ethernet networks.

codec Literally, COmpressor/DECompressor. A mathematical algorithm used to make a video or audio clip smaller, while maintaining a predetermined level of quality.

Common Internet File System (CIFS) Network protocol for a Windows-based file server.

composite clip Two files found within the same folder of a device: an RGB clip and its corresponding, separate alpha channel clip. Both files usually have the same name before their file extension.

contentbase A device used exclusively by Final Cut Server to store files for specific tasks like proxies and version control. Contentbases are formed as bundles to protect their contents from casual interaction in the Finder.

D

DNS (Domain Name System) A service provided on an Ethernet network that resolves IP address numbers (such as 192.168.1.20) to domain names (such as fcserver.metamediatech.com). A well-implemented DNS system provides these resolutions in both forward (name to number) and reverse (number to name) directions, and also passes questions it can't resolve to a larger DNS server elsewhere in the organization, or perhaps the Internet itself.

duplicate The process of copying the primary representation of an asset from one device to another (or back onto the same device, usually changing the name to differentiate it from the file that exists). The option to transcode the resulting file is available.

E

edit-in-place device A device that may be directly accessible to users using the client application. If it is, Final Cut Server can provide a path to a file, given in the form of a URI, rather than the file having to be downloaded to the user's cache.

edit proxy A representation of a video asset, used for the purposes of editing clips. Uses Apple's ProRes 422 codec.

export The process of downloading the primary representation of an asset to a specific location on your computer using the client application. You can opt to transcode the primary representation during this process, so that the file that is received by your computer is in a different format.

F

File Transfer Protocol (FTP) A common network protocol for a file server, used primarily for sharing files on the Internet. FTP servers are therefore commonly accessed from remote locations.

Final Cut Server The Mac that is running installed and configured Final Cut Server software. It may also refer to the interaction between the Final Cut Server and the client application in general.

Final Cut Server software The software that gets installed and configured on the Mac that will become the Final Cut Server.

G

group A collection of users. In Final Cut Server, groups have certain access and functionality privileges that are assigned to them through the association with a permission set.

H

Hierarchical Storage Management system (HSM) A storage system designed for backup and archive. HSMs can continuously accept files into their file system and then, according to predefined rules, transfer the data essence of the file to more stable storage media, like tape, even though the file will still appear to be on the original file system. When a user or process wants to restore the file, it will appear that the file is there, even though its data essence is being retrieved from tape.

L

Library A filesystem device, created by Final Cut Server during initial installation, intended to store commonly accessed files, such as stock media and logos.

lock The process that prevents an asset's representations or metadata from being modified or overwritten by any other user. Administrators can cancel locks.

lookups Another way of saying pop-up menus or pick lists. Lookups are assigned to metadata fields in order to control the choices that a user has in filling out a field.

M

metadata Data about data. A standardized model for including searchable information for a file.

metadata field The basic storage unit for a piece of metadata in Final Cut Server.

metadata group A collection of metadata fields, assembled to describe a feature of an asset or production. Specialized metadata groups are also used in Final Cut Server to describe the fields shown in search results, as well as for filtering criteria used during advanced searches.

metadata set A collection of metadata groups that defines all of the metadata for any asset, production, or job within Final Cut Server.

N

Network File System (NFS) A network file server protocol used in many Unix/Linux environments.

O

Open Directory Apple's marketing term for a collection of directory system binding and communication protocols, which includes its implementation of the open-source Lightweight Directory Access Protocol (LDAP), used as a centralized directory and authentication system. It also includes plug-ins to interact with other directory systems, including Active Directory.

P

permission set A combination of functionality and access settings for a group.

Prepare for Disconnected Use The same process as Add to Cache, with the addition of making an alias of the file located within /Users/username/Documents/Final Cut Server/Media Aliases/. This allows the user to use the primary representation of the asset even when the user is not connected to Final Cut Server.

primary representation The original file that an asset refers to. In most cases, this is the full-resolution clip or high-resolution image that resides on a device.

production A "virtual folder" within Final Cut Server, meant as a means to gather together media assets, project assets, and related document assets within a single container.

proxies device A contentbase device dedicated to storing clip proxies.

R

Redundant Array of Independent Disks (RAID) A system that combines a series of hard disk drives together in order to provide greater speed than a single drive. It may also employ a redundancy provision that allows one or more of the drives to fail without losing data.

representation A file to which an asset is associated. An asset is associated with at least a primary representation. Other representations can be thumbnails, clip proxies, edit proxies, and poster frames.

S

scheduled event or **schedule** A trigger that occurs in an absolute or periodic time frame. These triggers usually start scans that add and purge media. They also can execute maintenance tasks such as clearing logs.

Server Message Block (SMB) A network protocol for a Windows-based file server.

still sequence A collection of images that has some sort of iterative numbering system in the file-name. The collection usually describes a moving image, with each file being the next sequential frame.

subscription A trigger that looks for changes in the metadata of an asset, production, or job. When the changes occur, a subscription then activates any number of responses.

T

transcode The process of converting a file from one format into another, using a specific codec.

U

unlock The process of removing a lock from an asset, making it, its representations, and its metadata available for modification by others.

upload The process of copying files from your computer to Final Cut Server using the client application. Files can be uploaded either by dragging them into the client application's window or by selecting Upload from the Server pop-up menu.

V

Version device A contentbase device dedicated to storing older versions of an asset, used by the version control system in Final Cut Server.

W

watcher A folder that is constantly being "watched" by Final Cut Server for new or modified content, using a poll. Watchers are assigned responses by the administrator, including ones to copy and/or transcode the files placed inside a watcher to other devices.

X

Xsan The marketing term for Apple's Storage Area Network (SAN) system. Xsan allows up to 64 computers to share large volumes of data with real-time access to every file.

Xserve Apple's enterprise-class, rack-mountable computer, designed for server applications, such as web server, file server, DNS, and Open Directory. Xserves have field-replaceable redundant drive modules, power supplies, and motherboards. They take up only one rack space (1U) in a standard 19-inch rack.

Index